Environmental, Atmospheric and Soil Science

ABOUT THE BOOK

In science, the environment is the all the factors and conditions (biological, chemical and physical) in which an organism lives. So if we are studying fish, its environment is the water it lives in, the temperature and chemical makeup of the water, the type of climate of the lake or pond, how much rain it gets, average daily temperatures, what type of other fish live in the lake, presence of algae or plants, and the list goes on and on. environmental science is the field of science that studies the interactions of the physical, chemical, and biological components of the environment and also the relationships and effects of these components with the organisms in the environment. The field of environmental science can be divided into three main goals, which are to learn how the natural world works, to understand how we as humans interact with the environment, and also to determine how we affect the environment. The third goal of determining how humans affect the environment also includes finding ways to deal with these effects on the environment. Environmental science is also referred to as an interdisciplinary field because it incorporates information and ideas from multiple disciplines. Within the natural sciences, such fields as biology, chemistry, and geology are included in environmental science. When most people think of environmental science, they think of these natural science aspects, but what makes environmental science such a complex and broad field is that it also includes fields from the social sciences and the humanities.

ABOUT THE AUTHOR

Davinder Sehgal, M.Sc in Geographical Information Science from The Univeristy of Manchester is a Young scientist with a passion in research, a writer by occasion and she takes pride in educating the community through writing on every ocassion.Most recently, her work landed a position of writing, where she authors books on physical geography and enviroment. She did her Environmental Science from the College Of Earth and Environmental Sciences, Punjab University, Pakistan in 2008, She worked as Environmental Inspector at the goverment of Punjab.

Environmental, Atmospheric and Soil Science

DAVINDER SEHGAL

WESTBURY PUBLISHING LTD.
ENGLAND (UNITED KINGDOM)

Environmental, Atmospheric and Soil Science
Edited by: Davinder Sehgal
ISBN: 978-1-913806-59-0 (Hardback)

© 2021 Westbury Publishing Ltd.

Published by **Westbury Publishing Ltd.**
Address: 6-7, St. John Street, Mansfield,
Nottinghamshire, England, NG18 1QH
United Kingdom
Email: - info@westburypublishing.com
Website: - www.westburypublishing.com

British Library Cataloguing in Publication Data:
A catalogue record for this book is available from the British Library.

For more information regarding Westbury Publishing Ltd and its products, please visit the publisher's website- **www.westburypublishing.com**

Preface

In science, the environment is the all the factors and conditions (biological, chemical and physical) in which an organism lives. So if we are studying fish, its environment is the water it lives in, the temperature and chemical makeup of the water, the type of climate of the lake or pond, how much rain it gets, average daily temperatures, what type of other fish live in the lake, presence of algae or plants, and the list goes on and on.

environmental science is the field of science that studies the interactions of the physical, chemical, and biological components of the environment and also the relationships and effects of these components with the organisms in the environment. The field of environmental science can be divided into three main goals, which are to learn how the natural world works, to understand how we as humans interact with the environment, and also to determine how we affect the environment. The third goal of determining how humans affect the environment also includes finding ways to deal with these effects on the environment. Environmental science is also referred to as an interdisciplinary field because it incorporates information and ideas from multiple disciplines. Within the natural sciences, such fields as biology, chemistry, and geology are included in environmental science. When most people think of environmental science, they think of these natural science aspects, but what makes environmental science such a complex and broad field is that it also includes fields from the social sciences and the humanities.

The social science fields that are incorporated into environmental science include geography, economics, and political science. Philosophy and ethics are the two fields within the humanities that are also included in environmental science. By combining aspects of the natural sciences, social sciences, and the humanities, the field of environmental science can cover more concepts and also examine problems and topics from many different points of view.

Related areas of study include environmental studies and environmental engineering. Environmental studies incorporates more of the social sciences for understanding human relationships, perceptions and policies towards the environment. Environmental engineering focuses on design and technology for improving environmental quality in every aspect.

Environmental scientists work on subjects like the understanding of earth processes, evaluating alternative energysystems, pollution control and mitigation, natural resource management, and the effects of global climate change. Environmental issues almost always include an interaction of physical, chemical, and biological processes. Environmental scientists bring a systems approach to the analysis of environmental problems. Key elements of an effective environmental scientist include the ability to relate space, and time relationships as well as quantitative analysis.

Environmental science came alive as a substantive, active field of scientific investigation in the 1960s and 1970s driven by (a) the need for a multi-disciplinary approach to analyze complex environmental problems, (b) the arrival of substantive environmental laws requiring specific environmental protocols of investigation and (c) the growing public awareness of a need for action in addressing environmental problems. Events that spurred this development included the publication of Rachel Carson's landmark environmental book Silent Spring along with major environmental issues becoming very public, such as the 1969 Santa Barbara oil spill, and the Cuyahoga River of Cleveland, Ohio, "catching fire" (also in 1969), and helped increase the visibility of environmental issues and create this new field of study.

Environmental studies is a multidisciplinary academic field which systematically studies human interaction with the environment in the interests of solving complex problems. Environmental studies brings together the principles of the physical sciences, commerce/economics and social sciences so as to solve contemporary environmental problems. It is a broad field of study that includes the natural environment, the built environment, and the sets of relationships between them. The field encompasses study in basic principles of ecology and environmental science, as well as associated subjects such as ethics, geography, anthropology, policy, politics, law, economics, philosophy, sociology and social justice, planning, pollution control and natural resource management. There are also many degree programs in Environmental Studies including the Master of Environmental Studies and the Bachelor of Environmental Studies.

—Editor

Contents

1

Introduction

In science, the environment is the all the factors and conditions (biological, chemical and physical) in which an organism lives. So if we are studying fish, its environment is the water it lives in, the temperature and chemical makeup of the water, the type of climate of the lake or pond, how much rain it gets, average daily temperatures, what type of other fish live in the lake, presence of algae or plants, and the list goes on and on.

Environmental science is the field of science that studies the interactions of the physical, chemical, and biological components of the environment and also the relationships and effects of these components with the organisms in the environment. The field of environmental science can be divided into three main goals, which are to learn how the natural world works, to understand how we as humans interact with the environment, and also to determine how we affect the environment. The third goal of determining how humans affect the environment also includes finding ways to deal with these effects on the environment. Environmental science is also referred to as an interdisciplinary field because it incorporates information and ideas from multiple disciplines. Within the natural sciences, such fields as biology, chemistry, and geology are included in environmental science. When most people think of environmental science, they think of these natural science aspects, but what makes environmental science such a complex and broad field is that it also includes fields from the social sciences and the humanities.

The social science fields that are incorporated into environmental science include geography, economics, and political science. Philosophy and ethics are the two fields within the humanities that are also included in environmental science. By combining aspects of the natural sciences, social sciences, and the humanities, the field of environmental science can cover more concepts and also examine problems and topics from many different points of view.

Related areas of study include environmental studies and environmental engineering. Environmental studies incorporates more of the social sciences for understanding human relationships, perceptions and policies towards the environment. Environmental engineering focuses on design and technology for improving environmental quality in every aspect.

Environmental scientists work on subjects like the understanding of earth processes, evaluating alternative energysystems, pollution control and mitigation, natural resource management, and the effects of global climate change. Environmental issues almost always include an interaction of physical, chemical, and biological processes. Environmental scientists bring a systems approach to the analysis of environmental problems. Key elements of an effective environmental scientist include the ability to relate space, and time relationships as well as quantitative analysis.

Environmental science came alive as a substantive, active field of scientific investigation in the 1960s and 1970s driven by (a) the need for a multi-disciplinary approach to analyze complex environmental problems, (b) the arrival of substantive environmental laws requiring specific environmental protocols of investigation and (c) the growing public awareness of a need for action in addressing environmental problems. Events that spurred this development included the publication of Rachel Carson's landmark environmental book Silent Spring along with major environmental issues becoming very public, such as the 1969 Santa Barbara oil spill, and the Cuyahoga River of Cleveland, Ohio, "catching fire" (also in 1969), and helped increase the visibility of environmental issues and create this new field of study.

Environmental studies is a multidisciplinary academic field which systematically studies human interaction with the environment in the interests of solving complex problems. Environmental studies brings together the principles of the physical sciences, commerce/economics and social sciences so as to solve contemporary environmental problems. It is a broad field of study that includes the natural environment, the built environment, and the sets of relationships between them. The field encompasses study in basic principles of ecology and environmental science, as well as associated subjects such as ethics, geography, anthropology, policy, politics, law, economics, philosophy, sociology and social justice, planning, pollution control and natural resource management. There are also many degree programs in Environmental Studies including the Master of Environmental Studies and the Bachelor of Environmental Studies.

Environmental Protection Act (1986) defined "Environment as the sum total of water, air and land, their interrelationship among themselves and with the human beings, other living beings and property."

MEANING OF ENVIRONMENTAL STUDIES

Environmental studies are the scientific study of the environmental system and the status of its inherent or induced changes on organisms. It includes not only the study of physical and biological characters of the environment but also the social and cultural factors and the impact of man on environment.

DEFINITIONS

Environment is derived from the French word Environ which means to encircle orsurround.

Environment is sum total of water, air, and land, inter-relationships among themselves and also with the human beings, other living organisms and property. The above definitiongiven in Environment Act, 1986 clearly indicates that environment includes all the physical and biological surroundings and their interactions.

Objectives and Guiding Principles of Environmental Studies:

According to UNESCO (1971), the objectives of environmental studies are:

(a) Creating the awareness about environmental problems among people.
(b) Imparting basic knowledge about the environment and its allied problems.
(c) Developing an attitude of concern for the environment.
(d) Motivating public to participate in environment protection and environment improvement.
(e) Acquiring skills to help the concerned individuals in identifying and solving environmental problems.
(f) Striving to attain harmony with Nature.

According to UNESCO, the guiding principles of environmental education should be as follows:

(a) Environmental education should be compulsory, right from the primary up to the post graduate stage.
(b) Environmental education should have an interdisciplinary approach by including physical, chemical, biological as well as socio-cultural aspects of the environment. It should build a bridge between biology and technology.
(c) Environmental education should take into account the historical perspective, the current and the potential historical issues.

(d) Environmental education should emphasise the importance of sustainable development i.e., economic development without degrading the environment.

(e) Environmental education should emphasise the necessity of seeking international cooperation in environmental planning.

(f) Environmental education should lay more stress on practical activities and first hand experiences.

Scope and Importance of Environmental Studies:

The disciplines included in environmental education are environmental sciences, environmental engineering and environmental management.

(a) Environmental Science: It deals with the scientific study of environmental system (air, water, soil and land), the inherent or induced changes on organisms and the environmental damages incurred as a result of human interaction with the environment.

(b) Environmental Engineering: It deals with the study of technical processes involved in the protection of environment from the potentially deleterious effects of human activity and improving the environmental quality for the health and well beings of humans.

(c) Environmental Management: It promotes due regard for physical, social and economic environment of the enterprise or projects. It encourages planned investment at the start of the production chain rather than forced investment in cleaning up at the end.

It generally covers the areas as environment and enterprise objectives, scope, and structure of the environment, interaction of nature, society and the enterprise, environment impact assessment, economics of pollution, prevention, environmental management standards etc.

The importance's of environmental studies are as follows:

1. To clarify modern environmental concept like how to conserve biodiversity.
2. To know the more sustainable way of living.
3. To use natural resources more efficiently.
4. To know the behaviour of organism under natural conditions.
5. To know the interrelationship between organisms in populations and communities.
6. To aware and educate people regarding environmental issues and problems at local, national and international levels.

Need of Public Awareness about Environment:

In today's world because of industrialization and increasing population, the natural resources has been rapidly utilised and our environment is being increasingly degraded by human activities, so we need to protect the environment.

It is not only the duty of government but also the people to take active role for protecting the environment, so protecting our environment is economically more viable than cleaning it up once, it is damaged.

The role of mass media such as newspapers, radio, television, etc is also very important to make people aware regarding environment. There are various institutions, which are playing positive role towards environment to make people aware regarding environment like BSI (Botanical Survey of India, 1890), ZSI (Zoological Survey of India, 1916), WII (Wild Life Institute of India, 1982) etc.

ENVIRONMENTAL HISTORY

Environmental history is the study of human interaction with the natural world over time. In contrast to other historical disciplines, it emphasizes the active role nature plays in influencing human affairs. Environmental historians study how humans both shape their environment and are shaped by it.

Environmental history emerged in the United States out of the environmental movement of the 1960s and 1970s, and much of its impetus still stems from present-day global environmental concerns. The field was founded on conservation issues but has broadened in scope to include more general social and scientific history and may deal with cities, population or sustainable development. As all history occurs in the natural world, environmental history tends to focus on particular time-scales, geographic regions, or key themes. It is also a strongly multidisciplinary subject that draws widely on both the humanities and natural science.

The subject matter of environmental history can be divided into three main components. The first, nature itself and its change over time, includes the physical impact of humans on the Earth's land, water, atmosphere and biosphere. The second category, how humans use nature, includes the environmental consequences of increasing population, more effective technology and changing patterns of production and consumption. Other key themes are the transition from nomadic hunter-gatherer communities to settled agriculture in the neolithic revolution, the effects of colonial expansion and settlements, and the environmental and human consequences of the industrial and technological revolutions. Finally, environmental historians study how people think about nature - the way attitudes, beliefs and values influence interaction with nature, especially in the form of myths, religion and science.

Origin of name and early works

In 1967 Roderick Nash published "W ilderness and the American Mind", a work that has become a classic text of early environmental history. In an address to the Organization of American Historians in 1969 (published in 1970) Nash used the expression "environmental history", although 1972 is generally taken as the date when the term was first coined. The 1959 book by Samuel P. Hays, Conservation and the Gospel of Efficiency: The Progressive Conservation Movement, 1890-1920, while being a major contribution to American political history, is now also regarded as a founding document in the field of environmental history. Hays is Professor Emeritus of History at the University of Pittsburgh.

HISTORIOGRAPHY

Brief overviews of the field of environmental history have been given by John McNeill in 2003, Richard White in 1985, and J. Donald Hughes in 2006. In 2014 Oxford University Press published a volume of 25 essays entitled The Oxford Handbook of Environmental History. This collection was edited by Andrew C. Isenberg.

Definition

There is no universally accepted definition of environmental history. In general terms it is a history that tries to explain why our environment is like it is and how humanity has influenced its current condition, as well as commenting on the problems and opportunities of tomorrow. Donald Worster's widely quoted 1988 definition states: "Environmental history is the interaction between human cultures and the environment in the past."

In 2001 J. Donald Hughes defined the subject as "The study of human relationships through time with the natural communities of which they are a part in order to explain the processes of change that affect that relationship." and, in 2006, as "... history that seeks understanding of human beings as they have lived, worked and thought in relationship to the rest of nature through the changes brought by time" ... "As a method, environmental history is the use of ecological analysis as a means of understanding human history ... an account of changes in human societies as they relate to changes in the natural environment." Environmental historians are also "interested in what people think about nature, and how they have expressed those ideas in folk religions, popular culture, literature and art." In 2003 McNeill suggested that environmental history was "... the history of the mutual relations between humankind and the rest of nature".

Subject matter

Traditional historical analysis has over time extended its range of study from the activities and influence of a few significant people to a much broader social, political, economic and cultural analysis. Environmental history further broadens the subject matter of conventional history. In 1988, Donald Worster stated that environmental history "attempts to make history more inclusive in its narratives" by examining the "role and place of nature in human life", and in 1993, that "Environmental history explores the ways in which the biophysical world has influenced the course of human history and the ways in which people have thought about and tried to transform their surroundings". The interdependency of human and environmental factors in the creation of landscapes is expressed through the notion of the cultural landscape. Worster also questioned the scope of the discipline, asking: "W e study humans and nature; therefore can anything human or natural be outside our enquiry?"

Environmental history is generally treated as a subfield of history, an established discipline. But some environmental historians challenge this assumption, arguing that while traditional history is human history – the story of people and their institutions, "humans cannot place themselves outside the principles of nature." In this sense environmental history is a version of human history within a larger context, one less dependent on anthropocentrism (even though anthropogenic change is at the center of its narrative).

Dimensions

J. Donald Hughes responded to the view that environmental history is "light on theory" or lacking theoretical structure by viewing the subject through the lens of three "dimensions": nature and culture, history and science, and scale. This advances beyond Worster's recognition of three broad clusters of issues to be addressed by environmental historians although both historians recognize that the emphasis of their categories might vary according to the particular study as, clearly, some studies will concentrate more on society and human affairs and others more on the environment.

Themes

Several themes are used to express these historical dimensions. A more traditional historical approach is to analyse the transformation of the globe's ecology through themes like the separation of man from nature during the neolithic revolution, imperialism and colonial expansion, exploration, agricultural change, the effects of the industrial and technological revolution, and urban expansion. More environmental topics include human impact through influences on forestry, fire, climate change, sustainability and so on. According to Paul Warde, "the increasingly sophisticated history of colonization and migration can take on an environmental aspect, tracing the pathways of ideas and species around the globe and indeed is bringing about an increased use of such analogies and 'colonial' understandings of processes within European history." The importance of the colonial enterprise in Africa, the Caribbean and Indian Ocean has been detailed by Richard Grove. Much of the literature consists of case-studies targeted at the global, national and local levels.

Scale

Although environmental history can cover billions of years of history over the whole Earth, it can equally concern itself with local scales and brief time periods. Many environmental historians are occupied with local, regional and national histories. Some historians link their subject exclusively to the span of human history – "every time period in human history" while others include the period before human presence on Earth as a legitimate part of the discipline. Ian Simmons's Environmental History of Great Britain covers a period of about 10,000 years. There is a tendency to difference in time scales between natural and social phenomena: the causes of environmental change that stretch back in time may be dealt with socially over a comparatively brief period.

Although at all times environmental influences have extended beyond particular geographic regions and cultures, during the 20th and early 21st centuries anthropogenic environmental change has assumed global proportions, most prominently with climate change but also as a result of settlement, the spread of disease and the globalization of world trade.

History of the subject

The questions posed and themes covered by environmental history date back to antiquity: historians have always included the effects of natural phenomena on human affairs. Hippocrates, ancient Greek father of medicine, in his Airs, Waters, Places, asserted that different cultures and human temperaments could be related to the surroundings in which peoples lived.During the Enlightenment there was a rising awareness of the environment as concept and early environmental scientists addressed themes of sustainability via the subjects of natural history and medicine. However, the origins of the subject in its present form are generally traced to the twentieth century.

In 1929 a group of French historians founded the journal Annales, in many ways a forerunner of modern environmental history since it took as its subject matter the reciprocal global influences of the environment and human society. The idea of the impact of the physical environment on civilizations was espoused by this Annales School to describe the long term developments that shape human history by focusing away from political and intellectual history, toward agriculture, demography, and geography. Emmanuel Le Roy Ladurie, a pupil of the Annales School, was the first to really embrace, in the 1950s, environmental history in a more contemporary form. One of the most influential members of the Annales School was Lucien Febvre (1878– 1956), whose book A Geographical Introduction to History is now a classic in the field.

The most influential empirical and theoretical work in the subject has been done in the United States where teaching programs first emerged and a generation of trained environmental historians is now active. In the United States environmental history as an independent field of study emerged in the general cultural reassessment and reform of the 1960s and 1970s along with environmentalism, "conservation history", and a gathering awareness of the global scale of some environmental issues. This was in large part a reaction to the way nature was represented in history at the time, which

"portrayed the advance of culture and technology as releasing humans from dependence on the natural world and providing them with the means to manage it [and] celebrated human mastery over other forms of life and the natural environment, and expected technological improvement and economic growth to accelerate". Environmental historians intended to develop a post-colonial historiography that was "more inclusive in its narratives".

MORAL AND POLITICAL INSPIRATION

Moral and political inspiration to environmental historians has come from American writers and activists Henry Thoreau (1817 – 1862), John Muir (1838 – 1914), Aldo Leopold (1887 – 1948), and Rachel Carson (1907 – 1964). Environmental history frequently promoted a moral and political agenda although it steadily became a more scholarly enterprise." Early attempts to define the field were made in the United States by Roderick Nash in "The State of Environmental History" and in other works by frontier historians Frederick Jackson Turner, James Malin, and Walter Prescott Webb who analysed the process of settlement. Their work was expanded by a second generation of more specialized environmental historians such as Alfred

Crosby, Samuel P. Hays, Donald Worster, William Cronon, Richard White, Carolyn Merchant, John McNeill, Donald Hughes, Chad Montrie, and Europeans Paul Warde, Sverker Sorlin, Robert A. Lambert, T.C. Smout and Peter Coates.

British Empire

Although environmental history was growing rapidly after 1970, it only reached historians of the British Empire in the 1990s. Gregory Barton argues that the concept of environmentalism emerged from forestry studies, and emphasizes the British imperial role in that research. He argues that imperial forestry movement in India around 1900 included government reservations, new methods of fire protection, and attention to revenue-producing forest management. The result eased the fight between romantic preservationists and laissez-faire businessmen, thus giving the compromise from which modern environmentalism emerged.

In recent years numerous scholars cited by James Beattie have examined the environmental impact of the Empire. Beinart and Hughes argue that the discovery and commercial or scientific use of new plants was an important concern in the 18th and 19th centuries. The efficient use of rivers through dams and irrigation projects was an expensive but important method of raising agricultural productivity. Searching for more efficient ways of using natural resources, the British moved flora, fauna and commodities around the world, sometimes resulting in ecological disruption and radical environmental change. Imperialism also stimulated more modern attitudes toward nature and subsidized botany and agricultural research. Scholars have used the British Empire to examine the utility of the new concept of eco-cultural networks as a lens for examining interconnected, wide-ranging social and environmental processes.

Current practice

In the United States the American Society for Environmental History was founded in 1975 while the first institute devoted specifically to environmental history in Europe was established in 1991, based at the University of St. Andrews in Scotland. In 1986, the Dutch foundation for the history of environment and hygiene Net Werk was founded and publishes four newsletters per year. In the UK the White Horse Press in Cambridge has, since 1995, published the journal Environment and History which aims to bring scholars in the humanities and biological sciences closer together in constructing long and well-founded perspectives on present day environmental problems and a similar publication Tijdschrift voor Ecologische Geschiedenis (Journal for Environmental History) is a combined Flemish-

Dutch initiative mainly dealing with topics in the Netherlands and Belgium although it also has an interest in European environmental history. Each issue contains abstracts in English, French and German. In 1999 the Journal was converted into a yearbook for environmental history. In Canada the Network in Canadian History and Environment facilitates the growth of environmental history through numerous workshops and a significant digital infrastructure including their website and podcast.

Communication between European nations is restricted by language difficulties. In April 1999 a meeting was held in Germany to overcome these problems and to co-ordinate environmental history in Europe. This meeting resulted in the creation of the European Society for Environmental History in 1999. Only two years after its establishment, ESEH held its first international conference in St. Andrews, Scotland. Around 120 scholars attended the meeting and 105 papers were presented on topics covering the whole spectrum of environmental history. The conference showed that environmental history is a viable and lively field in Europe and since then ESEH has expanded to over 400 members and continues to grow and attracted international conferences in 2003 and 2005. In 1999 the Centre for Environmental History was established at the University of Stirling. Some history departments at European universities are now offering introductory courses in environmental history and postgraduate courses in Environmental history have been established at the Universities of Nottingham, Stirling and Dundee and more recently a Graduierten Kolleg was created at the University of Göttingen in Germany . In 2009, the Rachel Carson Center for Environment and Society (RCC), an international, interdisciplinary center for research and education in the environmental humanities and social sciences, was founded as a joint initiative of Munich's Ludwig-Maximilians-Universität and the Deutsches Museum, with the generous support of the German Federal Ministry of Education and Research. The Environment & Society Portal (environmentandsociety.org) is the Rachel Carson Center's open access digital archive and publication platform.

Importance of Environmental Science

At this current time, the world around us is changing at a very rapid pace. Some changes are beneficial, but many of the changes are causing damage to our planet. The field of environmental science is a valuable resource for learning more about these changes and how they affect the world we live in.

Let's examine a major change that is currently occurring and its relationship to environmental science. The large change is the dramatic increase in the number of humans on earth. For most of human history, the population has been less than a million people, but the current population has skyrocketed to over seven billion people. This equals out to seven thousand times more people!

Due to this increase in the human population, there has also been an increase in pressure on the natural resources and ecosystem services that we rely on for survival. Natural resources include a variety of substances and energy sources that we take from the environment and use. Natural resources can be divided into renewable and nonrenewable resources. Renewable natural resources are substances that can be replenished over a period of time, such as sunlight, wind, soil, and timber. On the other hand, nonrenewable natural resources are substances that are in finite supply and will run out. Nonrenewable resources include minerals and crude oils.

Due to the increase in the human population, natural resources are being used up at a more rapid rate than in the past. Although renewable natural resources can be replenished, when they are used too rapidly, they cannot be replenished fast enough to meet human demand. Even worse, when nonrenewable natural resources are used too rapidly, they become closer to running out completely and being gone forever.

Natural resources have been referred to as the 'merchandise' produced by the environment, and in this respect, ecosystem services are the 'facilities' that we rely on to help produce the merchandise. Ecosystemservices are the environment's natural processes that provide us with the resources we need to support life. Common ecosystem services include water and air purification, nutrient cycling, climate regulation, pollinatingof plants, and the recycling of waste. Just like some natural resources, ecosystem services are also limited and can be used up if not regulated.

Now, let's tie it together and think about population growth and its influence on both natural resources and ecosystem services. As the human population increases and natural resources and ecosystem services are used rapidly and potentially degraded, the future of humans on earth is in jeopardy. This is one major example of why environmental science is important and valuable.

COMPONENTS OF ENVIRONMENTAL SCIENCE

Ecology

Ecology is the study of organisms and the environment interacting with one another. Ecologists, who make up a part of environmental scientists, try to find relations between the status of the environment and the population of a particular species within that environment, and if there is any correlations to be drawn between the two. For example, ecologists might take the populations of a particular type of bird with the status of the part of the Amazon Rainforest that population is living in.

The ecologists will study and may or may not come to the conclusion that the bird population is increasing or decreasing as a result of air pollution in the rainforest. They may also take multiple species of birds and see if they can find any relation to one another, allowing the scientists to come to a conclusion if the habitat is suitable or not for that species to live in.

Geoscience

Geoscience concerns the study of geology, soil science, volcanoes, and the Earth's crust as they relate to the environment. As an example, scientists may study the erosion of the Earth's surface in a particular area. Soil scientists, physicists, biologists, and geomorphologists would all take part in the study.

Geomorphologists would study the movement of solid particles (sediments), biologists would study the impacts of the study to the plants and animals of the immediate environment, physicists would study the light transmission changes in the water causing the erosion, and the soil scientists would make the final calculations on the flow of the water when it infiltrates the soil to full capacity causing the erosion in the first place.

Atmospheric Science

Atmospheric science is the study of the Earth's atmosphere. It analyzes the relation of the Earth's atmosphere to the atmospheres of other systems. This encompasses a wide variety of scientific studies relating to space, astrology and the Earth's atmosphere: meteorology, pollution, gas emissions, and airborne contaminants.

An example of atmospheric science is where physicists study atmospheric circulation of a part of the atmosphere, chemists would study the chemicals existent in this part and their relationships with the environment, meteorologists study the dynamics of the atmosphere, and biologists study how the plants and animals are affected and their relationship with the environment.

Environmental Chemistry

Environmental Chemistry is the study of the changes chemicals make in the environment, such as contamination of the soil, pollution of the water, degradation of chemicals, and the transport of chemicals upon the plants and animals of the immediate environment. An example of environmental chemistry would be introduction of a chemical object into an environment, in which chemists would then study the chemical bonding to the soil or sand of the environment. Biologists would then study the now chemically induced soil to see its relationship with the plants and animals of the environment.

Environmental science is an active and growing part of the scientific world accelerated by the need to address problems with the Earth's environment. It encompasses multiple scientific fields and sciences to see how all interchange and relate with one another in any of the above four components.

Scope for Environmental Science

It is necessary to conserve the environment and save the mankind from awaiting natural disasters due to the indiscriminate exploitation, depletion of natural resources and accompanied by environmental pollution. Therefore, in the recent days, awareness on environmental protection and natural resources is slowly on rise.

Execution of pollution control and eradication of waste products can only be carried out through multi-pronged approach and a well-planned systematic research. The science of environment is a blend of Social Science and Nature.

Nature of job

Environment is a part of eco system. Environmental education is both valuable and necessary. Besides being an extensive field, it contains abundant job opportunities. Many features of this science are quite new to India. But, it is a prominent department in foreign countries. Numerous job opportunities are available like-Environmental Expert, Environmental Engineers, Environmental Creator and Environmental Journalist.

Requirements in Industries

Environmental scientists are concerned with work related to Biodiversity, Waste Land Management, Environmental Balance and Protection of Natural Resources. Hence, the environmental scientists get lot of opportunities to work in industries associated with environment. They are involved in work concerned with pollution control and waste management.

Research and Development

Several universities and government-based institutions offer opportunities to experimenting jobs for undertaking Researches and developments. They are involved in protecting the Natural Resources from the danger of Pollution. Forestry and Wild Life Management - As forests and wild life apart from enhancing the beauty of Earth also provide an eco-balance on the earth, hence the protection of forests and wild life is absolutely necessary at the moment.

Social Development

The Non Governmental Organisations (NGOs) take the problems concerning the environment to the public. They also try to create awareness on environmental protection. Population control is also emphasized in this.

Environmental Experts

These people focus on cleaning-up of lakes, Rivers and Forests. Apart from their prime job of growing forests and creating new forests, they are also involved in creating awareness.

Environmental Journalism

It is important to publish news regarding environment since environmental awareness happens amidst public. These journalists carry all news items regarding environment to the public through newspapers.

Counsellors

Private counsellors who work along with both the Central and the State Government Pollution Control Board, give right suggestions to the Government. They also provide legal information to industries regarding environmental protection. Some of them concentrate on Project Planning, Pollution Control, Eco- Balance and Protection.

Educational Qualification

To work as a scientist in this field, one has to complete Post graduation in Environmental Science or ME in Environmental Chemical Engineering. Those who have completed Bsc/Msc in Environmental Science are selected to administer the NGOs. Generally, this field of study is available as a Postgraduate programme.

Job Opportunities

In this field, several job opportunities are available in India. Great dearth of professionals prevails to write environmental articles for creating awareness among the public. Opportunities are available for scientists to get into Mining, Fertilizer Manufacturing Unit, Textile Industry, Dying Units and Food Processing Unit.

Moreover, there is a constant need for NGOs, and the NGOs are likely to provide Rs5, 000/- to 11,000/-pm.; the Researcher can earn about Rs.12, 000/- and a teacher can get about Rs 7,000 to 10,000 pm as salary.

SCOPE OF ENVIRONMENTAL SCIENCE IN INDIA

Environmental science is a subject which effects the whole world with it's work impact and nowadays it is taken seriously all over.

Now we talk about India, it is also working in the field to conserve and protect the nature and the biodiversity present here. Environmental science has a very large range of scope in India , but you have to find institutions to pursue the subject and you will find it rarely. People think it is a waste and timepass type of subject but it's not as it takes the responsibility to teach the student how to protect our environment and biodiversity , which I think is not a piece of cake. If you find a good college and get good knowledge about the subject you can get many jobs in India. Yes, I agree India has not given the respect to the subject it deserve but it is spreading all over the country and now we are aware of the environmental issues and seek the knowledge and care for it, and this is the point where people need environmentalists to solve their problems. The various job scopes are in industrial sectors, governmental field jobs or governmental surveys. Beside these you can be teacher and professor after getting so much knowledge as many college are offering environmental science as honors now and even now every college and school has seconday paper of the subject which needs teachers .

Environmental science is a wide and interesting subject which has responsibility to save the ecosystem and is getting light in every nation and has great worldwide scope.

Environmental Science Careers

1. Environmental Scientist: An environmental scientist performs research to pinpoint, minimize the grave impacts of, or get rid of hazards and pollutants to the environments or the health of the globe's population. His or her main aim is to protect and conserve Mother Nature. The average annual salary of an environmental scientist is $67, 400. Environmental scientist jobs are normally full time, including working over 40 hours a week. Their main workplaces are laboratories and outdoor field works. You need impeccable skills set to be an environmental scientist including communication skills, reading comprehension, and self-discipline.

2. Environmental Engineer: An environmental engineer finds solutions to problems in the environment by leveraging his or her knowledge of soil science, engineering, biology, and chemistry. His or her area of focus includes control of pollution, recycling, and Public health aspects. A bachelor's degree in environmental engineering is required

to become an environmental engineer. Environmental engineers take home an annual salary of about $83,360. An environmental engineer should be able to:

- Offer suggestions for maintaining and beefing up environmental performance
- Find out, evaluate and apply storm water good management practices for municipal, industrial and construction stormwater programs
- Evaluate environmental regulations and seek counsel with applicability determination
- He or she should document all environmental incidences
- Develop and keep in line environmental management systems to conform to air and permit regulations.
- Lead from the front in the negotiation and of permit applications
- Liaise with regulatory bodies, prepare required documentation, organize any testing sessions and provide more follow-up documentation needed.

3. Environmental Biologist: An environmental biologist deals with ecosystem and wildlife that reside in it. However, environmental biologist focuses a lot on the biological side of any ecosystem, which means his or her duties are more inclined to biology. To qualify as an environmental biologist, you need a bachelor's degree in environmental biology. The duties of an environmental biologist include:

- Administering biological and project schedules using scientific techniques, statistical tools and in-depth knowledge in the domain
- Scheduling and conducting biological experiments and follow up with field and laboratory operating processes
- Pinpoint project technicalities and counteract them effectively without compromising resources and time
- Bring up findings of experiments to colleagues and stakeholders
- Ensure scientific integrity by working with a team in conjunction with peer review data

4. Environmental Geologist: Environmental geology is a field of study that blends the main cornerstone of environmental science and puts more emphasis on geology and how to apply it in real world scenario to solve environmental problems. An environmental geologist goes beyond interest in environment and correlation between humans and wildlife in it. He or she also focuses on fossil fuels, non-renewable resources, minerals and the earth's crust.

An environmental geologist will spend most of his or her work life serving as a consultant, assisting in the mitigation of soil and ground water contamination by determining the right location for new landfills and drawing up a plan for underground waste disposal. He or she also ensures water supplies and soils that are responsible for growing food stay clean and uncontaminated without affecting or limiting accessibility to any organism or animal that relies on them for survival. The average annual salary for an environmental geologist is $89 700.

5. Environmental Biotechnologist: This professional synchronizes engineering and biology to create and utilize procedures that correct contaminated sites. For instance, there is a wide range of bacteria, microbes, and fungi that are able to consume pollutants and later break them down into safe elements over time. An environmental biotechnologist finds out, utilize and create necessary microbes for correcting a specific area and the pollutants unique to it. Contaminated soil might be corrected or remedied on site or loaded into containers and transported for treatment. Environmental biotechnologists perform the following duties:

- Develop plant-based bioplastics
- Transform plants into biofuels
- Draw up correction plans for specific sites that abide by environmental regulations
- Develop procedures to convert waste into biogas or other cleaner sources of energy
- Develops cleaner industrial processes by substituting chemicals with biological processes
- Utilize GIS (Geographic Information Systems) to pinpoint contaminated sites and distribution of pollutants

Typically, one needs to have master's and doctorate degrees to qualify as an environmental biotechnologist. However, the exponential growth of the domain has opened up a whole lot of opportunities for those having bachelor's and associate's degrees.

Regulations driving the studies

In the U.S. the National Environmental Policy Act (NEPA) of 1969 set forth requirements for analysis of major projects in terms of specific environmental criteria. Numerous state laws have echoed these mandates, applying the principles to local-scale actions. The upshot has been an explosion of documentation and study of environmental consequences before the fact of development actions.

One can examine the specifics of environmental science by reading examples of Environmental Impact Statements prepared under NEPA such as: Wastewater treatment expansion options discharging into the San Diego/ Tijuana Estuary, Expansion of the San Francisco International Airport, Development of the Houston, Metro Transportation system, Expansion of the metropolitan Boston MBTA transit system, and Construction of Interstate 66 through Arlington, Virginia.

In England and Wales the Environment Agency (EA), formed in 1996, is a public body for protecting and improving the environment and enforces the regulations listed on the communities and local government site. (formerly the office of the deputy prime minister). The agency was set up under the Environment Act 1995 as an independent body and works closely with UK Government to enforce the regulations.

2

Environmental Chemistry

Environmental chemistry is a very focused branch of chemistry, containing aspects of organic chemistry, analytical chemistry, physical chemistry and inorganic chemistry, as well as more diverse areas, such as biology, toxicology, biochemistry, public health and epidemiology. Environmental chemists work in a variety of public, private and government laboratories.

Environmental chemistry is socially important because it deals with the environmental impact of pollutants, the reduction of contamination and management of the environment. Environmental chemistists study the behavior of pollutants and their environmental effects on the air, water and soil environments, as well as their effects on human health and the natural environment.

Environmental chemistry is more than just water, air, soil, and chemicals. People in this field use math, biology, genetics, hydrology, engineering, toxicology, and a lot more to help answer important questions about our environment, the chemicals therein, and what role people play in all of this or how it impacts us as a result.

Environmental chemistry is the study of chemical processes occurring in the environment which are impacted by humankind's activities. These impacts may be felt on a local scale, through the presence of urban air pollutants or toxic substances arising from a chemical waste site, or on a global scale, through depletion of stratospheric ozone or global warming. The focus in our courses and research activities is upon developing a fundamental understanding of the nature of these chemical processes, so that humankind's activities can be accurately evaluated.

Environmental chemistry involves first understanding how the uncontaminated environment works, which chemicals in what concentrations are present naturally, and with what effects. Without this it would be impossible to accurately study the effects humans have on the environment through the release of chemicals.

Environmental chemists draw on a range of concepts from chemistry and various environmental sciences to assist in their study of what is happening to a chemical species in the environment. Important general concepts from chemistry include understanding chemical reactions and equations, solutions, units, sampling, and analytical techniques.

CONTAMINATION

A contaminant is a substance present in nature at a level higher than fixed levels or that would not otherwise be there. This may be due to human activity and bioactivity. The term contaminant is often used interchangeably with pollutant, which is a substance that has a detrimental impact on the surrounding environment.Whilst a contaminant is sometimes defined as a substance present in the environment as a result of human activity, but without harmful effects, it is sometimes the case that toxic or harmful effects from contamination only become apparent at a later date.

The "medium" (e.g. soil) or organism (e.g. fish) affected by the pollutant or contaminant is called a receptor, whilst a sink is a chemical medium or species that retains and interacts with the pollutant e.g. as carbon sink and its effects by microbes.

Freshwater environmental quality parameters

Freshwater environmental quality parameters are the natural and man-made chemical, biological and microbiological characteristics of rivers, lakes and ground-waters, the ways they are measured and the ways that they change. The values or concentrations attributed to such parameters can be used to describe the pollution status of an environment, its biotic status or to predict the likelihood or otherwise of a particular organisms being present. Monitoring of environ Green chemistry mental quality parameters is a key activity in managing the environment, restoring polluted environments and anticipating the effects of man-made changes on the environment.

Freshwater environmental quality parameters are those chemical, physical or biological parameters that can be used to characterise a freshwater body. Because almost all water bodies are dynamic in their composition, the relevant quality parameters are typically expressed as a range of expected concentrations.

Characterisation

The first step in understanding the chemistry of freshwater is to establish the relevant concentrations of the parameters of interest. Conventionally this is done by taking representative samples of the water for subsequent analysis in a laboratory . However, in-situ monitoring using hand-held analytical equipment or using bank-side monitoring stations are also used.

Sampling

Freshwaters are surprisingly difficult to sample because they are rarely homogeneous and their quality varies during the day and during the year. In addition the most representative sampling locations are often at a distance from the shore or bank increasing the logistic complexity.

Rivers

Filling a clean bottle with river water is a very simple task, but a single sample is only representative of that point along the river the sample was taken from and at that point in time. Understanding the chemistry of a whole river, or even a significant tributary, requires prior investigation to understand how homogeneous or mixed the flow is and to determine if the quality changes during the course of a day and during the course of a year. Almost all natural rivers will have very significant patterns of change through the day and through the seasons. Water remote sensing offers a spatially continuous tool to improve understanding of spatial and temporal river water quality. Many rivers also have a very large flow that is unseen. This flows through underlying gravel and sand layers and is called hyporheic flow. How much mixing there is between the hyporheic zone and the water in the open channel will depend on a variety of factors, some of which relate to flows leaving aquifers which may have been storing water for many years.

Ground-waters

Ground waters by their very nature are often very difficult to access to take a sample. As a consequence the majority of ground-water data comes from samples taken from springs, wells, water supply bore-holes and in natural caves. In recent decades as the need to understand ground water dynamics has increased, an increasing number or monitoring bore-holes have been drilled into aquifers

Lakes

Lakes and ponds can be very large and support a complex eco-system in which environmental parameters vary widely in all three physical dimensions and with time. Large lakes in the temperate zone often stratify in the warmer months into a warmer upper layers rich in oxygen and a colder lower layer with low oxygen levels. In the autumn, falling temperatures and occasional high winds result in the mixing of the two layers into a more homogeneous whole. When stratification occurs it not only affects oxygen levels but also many related parameters such as iron, phosphate and manganese which are all changed in their chemical form by change in the redoxpotential of the environment.

Lakes also receive waters, often from many different sources with varying qualities. Solids from stream inputs will typically settle near the mouth of the stream and depending on a variety of factors the incoming water may float over the surface of the lake, sink beneath the surface or rapidly mix with the lake water. All of these phenomena can skew the results of any environmental monitoring unless the process are well understood.

Mixing zones

Where two rivers meet at a confluence there exists a mixing zone. A mixing zone may be very large and extend for many miles as in the case of the Mississippi and Missouri rivers in the United States and the River Clwyd and River Elwy in North Wales. In a mixing zone water chemistry may be very variable and can be difficult to predict. The chemical interactions arc not just simple mixing but may be complicated by biological processes from submerged macrophytes and by water joining the channel from the hyporheic zone or from springs draining an aquifer.

Geological inputs

The geology that underlies a river or lake has a major impact on its chemistry. A river flowing across very ancient precambrian schists is likely to have dissolved very little from the rocks and maybe similar to de-ionised water at least in the headwaters. Conversely a river flowing through chalk hills, and especially if its source is in the chalk, will have a high concentration of carbonates and bicarbonates of Calcium and possibly Magnesium.

As a river progresses along its course it may pass through a variety of geological types and it may have inputs from aquifers that do not appear on the surface anywhere in the locality.

Atmospheric inputs

Oxygen is probably the most important chemical constituent of surface water chemistry, as all aerobic organisms require it for survival. It enters the water mostly via diffusion at the water-air interface. Oxygen's solubility in water decreases as water temperature increases. Fast, turbulent streams expose more of the water's surface area to the air and tend to have low temperatures and thus more oxygen than slow, backwaters. Oxygen is a by-product of photosynthesis, so systems with a high abundance of aquatic algae and plants may also have high concentrations of oxygen during the day. These levels can decrease significantly during the night when primary producers switch to respiration. Oxygen can be limiting if circulation between the surface and deeper layers is poor, if the activity of animals is very high, or if there is a large amount of organic decay occurring such as following Autumn leaf-fall.

Most other atmospheric inputs come from man-made or anthropogenic sources the most significant of which are the oxides of sulphur produced by burning sulphur rich fuels such as coal and oil which give rise to acid rain. The chemistry of sulphur oxides is complex both in the atmosphere and in river systems. However the effect on the overall chemistry is simple in that it reduces the pH of the water making it more acidic. The pH change is most marked in rivers with very low concentrations of dissolved salts as these cannot buffer the effects of the acid input. Rivers downstream of major industrial conurbations are also at greatest risk. In parts of Scandinaviaand West Wales and Scotland many rivers became so acidic from oxides of sulphur that most fish life was destroyed and pHs as low as pH4 were recorded during critical weather conditions.

Anthropogenic inputs

The majority of rivers on the planet and many lakes have received or are receiving inputs from human-kind's activities. In the industrialised world, many rivers have been very seriously polluted, at least during the 19th and the first half of the 20th centuries. Although in general there has been much improvement in the developed world, there is still a great deal of river pollution apparent on the planet.

Toxicity

In most environmental situations the presence or absence of an organism is determined by a complex web of interactions only some of which will be related to measurable chemical or biological parameters. Flow rate, turbulence, inter and intra specific competition, feeding behaviour, disease, parasatism, commensalism and symbiosis are just a few of the pressures and opportunities facing any organism or population. Most chemical constituents favour some organisms and are less favourable to others. However, there are some cases where a chemical constituent exerts a toxic effect. i.e. where the concentration can kill or severely inhibit the normal functioning of the organism. Where a toxic effect has been demonstrated this may be noted in the sections below dealing with the individual parameters.

CHEMICAL CONSTITUENTS

Colour and turbidity

Often it is the colour of freshwater or how clear or hazy the water is that is the most obvious visual characteristic. Unfortunately neither colour nor turbidity are strong indicators of the overall chemical composition of water. However both colour and turbidity reduce the amount of light penetrating the water and can have significant impact on algae and macrophytes. Some algae in particular are highly dependent on water with low colour and turbidity

Many rivers draining high moor-lands overlain by peat have a very deep yellow brown colour caused by dissolved humic acids.

Organic constituents

One of the principal sources of elevated concentrations of organic chemical constituents is from treated sewage.

Dissolved organic material is most commonly measured using either the Biochemical oxygen demand (BOD) test or the Chemical oxygen demand (COD) test. Organic constituents are significant in river chemistry for the effect that they have on dissolved oxygen concentration and for the impact that individual organic species may have directly on aquatic biota.

Any organic and degradable material consumes oxygen as it decomposes. Where organic concentrations are significantly elevated the effects on oxygen concentrations can be significant and as conditions get extreme the river bed may become anoxic.

Some organic constituents such as synthetic hormones, pesticides, phthalates have direct metabolic effects on aquatic biota and even on humans drinking water taken from the river. Understanding such constituents and how they can be identified and quantified is becoming of increasing importance in the understanding of freshwater chemistry.

Metals

A wide range of metals may be found in rivers from natural sources where metal ores are present in the rocks over which the river flows or in the aquifers feeding water into the river. However many rivers have an increased load of metals because of industrial activities which include mining and quarrying and the processing and use of metals.

Iron

Iron, usually as Fe is a common constituent of river waters at very low levels. Higher iron concentrations in acidic springs or an anoxic hyporheic zone may cause visible orange/brown staining or semi-gelatinous precipitates of dense orange iron bacterial floc carpeting the river bed. Such conditions are very deleterious to most organisms and can cause serious damage in a river system.

Coal mining is also a very significant source of Iron both in mine-waters and from stocking yards of coal and from coal processing. Long abandoned mines can be a highly intractable source of high concentrations of Iron. Low levels of iron are common in spring waters emanating from deep-seated aquifers and maybe regarding as health giving springs. Such springs are commonly called Chalybeate springs and have given rise to a number of Spa towns in Europe and the United States.

Zinc

Zinc is normally associated with metal mining, especially Lead and Silver mining but is also a component pollutant associated with a variety of other metal mining activities and with Coal mining. Zinc is toxic at relatively low concentrations to many aquatic organisms. Microregma starts to show a toxic reaction at concentrations as low as 0.33 mg/l

Heavy metals

Lead and silver in river waters are commonly found together and associated with lead mining. Impacts from very old mines can be very long-lived. In the River Ystwythin Wales for example, the effects of silver and lead mining in the 17th and 18th centuries in the headwaters still causes unacceptably high levels of Zinc and Lead in the river water right down to its confluence with the sea. Silver is very toxic even at very low concentrations but leaves no visible evidence of its contamination.

Lead is also highly toxic to freshwater organisms and to humans if the water is used as drinking water. As with Silver, Lead pollution is not visible to the naked eye. The River Rheidol in west Wales had a major series of lead mines in its headwaters until the end of the 19th century and its mine discharges and waste tips remain to this day. In 1919 - 1921 only 14 species of invertebrates were found in the lower Rheidol when Lead concentrations were between 0.2ppm and 0.5ppm. By 1932 the lead concentration had reduced to 0.02ppm to 0.1ppm because of the abandonment of mining and, at those concentrations, the bottom fauna had stabilized to 103 species including three leeches.

Coal mining is also a very significant source of metals, especially Iron, Zinc and Nickel particularly where the coal is rich if pyrites which oxidises on contact with the air producing a very acidic leachate which is able to dissolve metals from the coal.

Significant levels of copper are unusual in rivers and where it does it occur the source is most likely to be mining activities, coal stocking, or pig farming. Rarely elevated levels may be of geological origin. Copper is acutely toxic to many freshwater organisms, especially algae, at very low concentrations and significant concentration in river water may have serious adverse effects on the local ecology.

Nitrogen

Nitrogenous compounds have a variety of sources including washout of oxides of nitrogen from the atmosphere, some geological inputs and some from macrophyte and algal nitrogen fixation. However, for many rivers in the proximity of humans, the largest input is from sewage whether treated or untreated. The nitrogen derives from breakdown products of proteins found in urine and faeces. These products, being very soluble, often pass through sewage treatment process and are discharged into rivers as a component of sewage treatment effluent. Nitrogen may be in the form of nitrate, nitrite, ammonia or ammonium salts or what is termed albuminoid nitrogen or nitrogen still within an organic proteinoid molecule.

The differing forms of nitrogen are relatively stable in most river systems with nitrite slowly transforming into nitrate in well oxygenated rivers and ammonia transforming into nitrite/ nitrate. However, the process are slow in cool rivers and reduction in concentration may more often be attributed to simple dilution. All forms of nitrogen are taken up by macrophytes and algae and elevated levels of nitrogen are often associated with overgrowths of plants or eutrophication. These can have the effect of blocking channels and inhibiting navigation. However, ecologically, the more significant effect is on dissolved oxygen concentrations which may become super-saturated during daylight due to plant photosynthesis but then drop to very low levels during darkness as plant respiration uses up the dissolved oxygen. Coupled with the release of oxygen in photosynthesis is the creation of bi-carbonate ions which cause a steep rise in pH and this is matched in darkness as carbon dioxide is released through respiration which substantially lowers the pH. Thus high levels of nitrogenous compounds tends to lead to eutrophication with extreme variations in parameters which in turn can substantially degrade the ecological worth of the watercourse.

Ammonium ions also have a toxic effect, especially on fish. The toxicity of ammonia is dependent on both pH and temperature and an added complexity is the buffering effect of the blood/water interface across the gill membrane which masks any additional toxicity over about pH 8.0. The management of river chemistry to avoid ecological damage is particularly difficult in the case of ammonia as a wide range of potential scenarios of concentration, pH and temperature have to be considered and the diurnal pH fluctuation caused by photosynthesis considered. On warm summer days with high-bi-carbonate concentrations unexpectedly toxic conditions can be created.

Phosphorus

Phosphorus compounds are usually found as relatively insoluble phosphates in river water and, except in some exceptional circumstances, their origin is agriculture or human sewage. Phosphorus can encourage excessive growths of plants and algae and contribute to eutrophication. If a river discharges into a lake or reservoir phosphate can be mobilised year after year by natural processes. In the summer time, lakes stratify so that warm oxygen rich water floats on top of cold oxygen poor water. In the warm upper layers - the epilimnion- plants consume the available phosphate. As the plants die in the late summer they fall into the cool water layers underneath - the hypolimnion - and decompose. During winter turn-over, when a lake becomes fully mixed through the action of winds on a cooling body of water - the phosphates are spread throughout the lake again to feed a new generation of plants. This process is one of the principal causes of persistent algal blooms at some lakes.

Arsenic

Geological deposits of arsenic may be released into rivers where deep ground-waters are exploited as in parts of Pakistan. Many metalloid ores such as lead, gold and copper contain traces of arsenic and poorly stored tailings may result in arsenic entering the hydrological cycle.

Solids

Inert solids are produced in all montane rivers as the energy of the water helps grind away rocks into gravel, sand and finer material. Much of this settles very quickly and provides an important substrate for many aquatic organisms. Many salmonid fish require beds of gravel and sand in which to lay their eggs. Many other types of solids from agriculture, mining, quarrying, urban run-off and sewage may block-out sunlight from the river and may block interstices in gravel beds making them useless for spawning and supporting insect life.

BACTERIAL, VIRAL AND PARASITE INPUTS

Both agriculture and sewage treatment produce inputs into rivers with very high concentrations of bacteria and viruses including a wide range of pathogenic organisms. Even in areas with little human activity significant levels of bacteria and viruses can be detected originating from fish and aquatic mammals and from animals grazing near rivers such as deer. Upland waters draining areas frequented by sheep, goats or deer may also harbour a variety of opportunistic human parasites such as liver fluke. Consequently, there are very few rivers from which the water is safe to drink without some form of sterilisation or disinfection. In rivers used for contact recreation such as swimming, safe levels of bacteria and viruses can be established based on risk assessment.

Under certain conditions bacteria can colonise freshwaters occasionally making large rafts of filamentous mats known as sewage fungus – usually Sphaerotilus natans. The presence of such organisms is almost always an indicator of extreme organic pollution and would be expected to be matched with low dissolved oxygen concentrations and high BOD vales.

E. coli bacteria have been commonly found in recreational waters and their presence is used to indicate the presence of recent faecal contamination, but E. coli presence may not be indicative of human waste. E. coli are found in all warm-blooded animals. E. coli have also been found in fish and turtles. Enterobacteria may also persist in the environment in mud, sediments, sand and soil for considerable lengths of time.

pH

pH in rivers is affected by the geology of the water source, atmospheric inputs and a range of other chemical contaminants. pH is only likely to become an issue on very poorly buffered upland rivers where atmospheric sulphur and nitrogen oxides may very significantly depress the pH as low as pH4 or in eutrophic alkaline rivers where photosynthetic bi-carbonate ion production in photosynthesis may drive the pH up above pH10

Applications

Environmental chemistry is used by the Environment Agency (in England and Wales), the United States Environmental Protection Agency, the Association of Public Analysts, and other environmental agencies and research bodies around the world to detect and identify the nature and source of pollutants. These can include:

- Heavy metal contamination of land by industry. These can then be transported into water bodies and be taken up by living organisms.
- Nutrients leaching from agricultural land into water courses, which can lead to algal blooms and eutrophication.
- Urban runoff of pollutants washing off impervious surfaces (roads, parking lots, and rooftops) during rain storms. Typical pollutants include gasoline, motor oil and other hydrocarbon compounds, metals, nutrients and sediment (soil).

• Organometallic compounds.

Methods

Quantitative chemical analysis is a key part of environmental chemistry, since it provides the data that frame most environmental studies.

Common analytical techniques used for quantitative determinations in environmental chemistry include classical wet chemistry, such as gravimetric, titrimetric and electrochemical methods. More sophisticated approaches are used in the determination of trace metals and organic compounds. Metals are commonly measured by atomic spectroscopy and mass spectrometry: Atomic Absorption Spectrophotometry (AAS) and Inductively Coupled Plasma Atomic Emission (ICP-AES) or Inductively Coupled Plasma Mass Spectrometric (ICP-MS) techniques. Organic compounds are commonly measured also using mass spectrometric methods, such as Gas chromatography-mass spectrometry (GC/MS) and Liquid chromatography-mass spectrometry (LC/MS). Tandem Mass spectrometry MS/MS and High Resolution/Accurate Mass spectrometry HR/AM offer sub part per trillion detection. Non-MS methods using GCs and LCs having universal or specific detectors are still staples in the arsenal of available analytical tools.

Other parameters often measured in environmental chemistry are radiochemicals. These are pollutants which emit radioactive materials, such as alpha and beta particles, posing danger to human health and the environment. Particle counters and Scintillation counters are most commonly used for these measurements. Bioassays and immunoassays are utilized for toxicity evaluations of chemical effects on various organisms. Polymerase Chain Reaction PCR is able to identify species of bacteria and other organisms through specific DNA and RNA gene isolation and amplification and is showing promise as a valuable technique for identifying environmental microbial contamination.

Published analytical methods

Peer-reviewed test methods have been published by government agencies and private research organizations. Approved published methods must be used when testing to demonstrate compliance with regulatory requirements.

Environmental monitoring

Environmental monitoring describes the processes and activities that need to take place to characterise and monitor the quality of the environment. Environmental monitoring is used in the preparation of environmental impact assessments, as well as in many circumstances in which human activities carry a risk of harmful effects on the natural environment. All monitoring strategies and programmes have reasons and justifications which are often designed to establish the current status of an environment or to establish trends in environmental parameters. In all cases the results of monitoring will be reviewed, analysed statistically and published. The design of a monitoring programme must therefore have regard to the final use of the data before monitoring starts.

Air quality monitoring

Air quality monitoring is performed using specialized equipment and analytical methods used to establish air pollutant concentrations.

Air monitors are operated by citizens, regulatory agencies, and researchers to investigate air quality and the effects of air pollution.

Interpretation of ambient air monitoring data often involves a consideration of the spatial and temporal representativeness of the data gathered, and the health effects associated with exposure to the monitored levels.

Since air pollution is carried by the wind, consideration of anemometer data in the area between sources and the monitor often provides insights on the source of the air contaminants recorded by an air pollution monitor.

Close to the earth's surface, the atmosphere normally gets colder with height, but on certain days, the atmosphere begins to get warmer with height a short distance from the earth's surface, and air emissions build up under this "cap" on the vertical mixing.

Topographic features (such as a valley) that prevent lateral atmospheric mixing, coupled with the vertical cap on atmospheric mixing caused by an inversion, can lead to especially high air pollutant concentrations, for example, the 1948 Donora smog.

Air dispersion models that combine topographic, emissions and meteorological data to predict air pollutant concentrations are often helpful in interpreting air monitoring data.

If an air monitor produces concentrations of multiple chemical compounds, a unique "chemical fingerprint" of a particular air pollution source may emerge from analysis of the data.

Soil monitoring

Soil monitoring is the process of collection of soil and testing in laboratory by analytical methods.

Soil sampling are of two types:

● Grab Sampling: in this method, sample is collected randomly from field

- Composite Sampling: In this method, mixing of multiple sub samples for larger and non-uniform fields.

In laboratory, soil can be tested for pH, Chlorides, Sulphates, Phosphates and other metals.

WATER QUALITY MONITORING

Design of environmental monitoring programmes

Water quality monitoring is of little use without a clear and unambiguous definition of the reasons for the monitoring and the objectives that it will satisfy. Almost all monitoring (except perhaps remote sensing) is in some part invasive of the environment under study and extensive and poorly planned monitoring carries a risk of damage to the environment. This may be a critical consideration in wilderness areas or when monitoring very rare organisms or those that are averse to human presence. Some monitoring techniques, such as gill netting fish to estimate populations, can be very damaging, at least to the local population and can also degrade public trust in scientists carrying out the monitoring.

Almost all mainstream environmentalism monitoring projects form part of an overall monitoring strategy or research field, and these field and strategies are themselves derived from the high levels objectives or aspirations of an organisation. Unless individual monitoring projects fit into a wider strategic framework, the results are unlikely to be published and the environmental understanding produced by the monitoring will be lost.

PARAMETERS

Chemical

Analyzing water samples for pesticides

The range of chemical parameters that have the potential to affect any ecosystem is very large and in all monitoring programmes it is necessary to target a suite of parameters based on local knowledge and past practice for an initial review. The list can be expanded or reduced based on developing knowledge and the outcome of the initial surveys.

Freshwater environments have been extensively studied for many years and there is a robust understanding of the interactions between chemistry and the environment across much of the world. However, as new materials are developed and new pressures come to bear, revisions to monitoring programmes will be required. In the last 20 years acid rain, synthetic hormoneanalogues, halogenated hydrocarbons, greenhouse gases and many others have required changes to monitoring strategies.

Biological

In ecological monitoring, the monitoring strategy and effort is directed at the plants and animals in the environment under review and is specific to each individual study.

However, in more generalised environmental monitoring, many animals act as robust indicators of the quality of the environment that they are experiencing or have experienced in the recent past. One of the most familiar examples is the monitoring of numbers of Salmonid fish such as brown trout or Atlantic salmon in river systems and lakes to detect slow trends in adverse environmental effects. The steep decline in salmonid fish populations was one of the early indications of the problem that later became known as acid rain.

In recent years much more attention has been given to a more holistic approach in which the ecosystem health is assessed and used as the monitoring tool itself. It is this approach that underpins the monitoring protocols of the Water Framework Directive in the European Union.

Radiological

Radiation monitoring involves the measurement of radiation dose or radionuclide contamination for reasons related to the assessment or control of exposure to ionizing radiation or radioactive substances, and the interpretation of the results. The 'measurement' of dose often means the measurement of a dose equivalent quantity as a proxy (i.e. substitute) for a dose quantity that cannot be measured directly. Also, sampling may be involved as a preliminary step to measurement of the content of radionuclides in environmental media. The methodological and technical details of the design and operation of monitoring programmes and systems for different radionuclides, environmental media and types of facility are given in IAEA Safety Guide RS–G-1.8 and in IAEA Safety Report No. 64.

Radiation monitoring is often carried out using networks of fixed and deployable sensors such as the US Environmental Protection Agency's Radnet and the SPEEDI network in Japan. Airborne surveys are also made by organizations like the Nuclear Emergency Support Team.

Microbiological

Bacteria and viruses are the most commonly monitored groups of microbiological organisms and even these are only of great relevance where water in the aquatic environment is subsequently used as drinking water or where water contact recreation such as swimming or canoeing is practised.

Although pathogens are the primary focus of attention, the principal monitoring effort is almost always directed at much more common indicator species such as Escherichia coli, supplemented by overall coliform bacteria counts. The rationale behind this monitoring strategy is that most human pathogens originate from other humans via the sewage strea therefore discharge an effluent which, although having a clean appearance, still contains many millions of bacteria per litre, the majority of which are relatively harmless coliform bacteria. Counting the number of harmless (or less harmful) sewage bacteria allows a judgement to be made about the probability of significant numbers of pathogenic bacteria or viruses being present. Where E. coli or coliform levels exceed pre-set trigger values, more intensive monitoring including specific monitoring for pathogenic species is then initiated.

Populations

Monitoring strategies can produce misleading answers when relaying on counts of species or presence or absence of particular organisms if there is no regard to population size. Understanding the populations dynamics of an organism being monitored is critical.

As an example if presence or absence of a particular organism within a 10 km square is the measure adopted by a monitoring strategy, then a reduction of population from 10,000 per square to 10 per square will go unnoticed despite the very significant impact experienced by the organism.

Monitoring programmes

All scientifically reliable environmental monitoring is performed in line with a published programme. The programme may include the overall objectives of the organisation, references to the specific strategies that helps deliver the objective and details of specific projects or tasks within those strategies. However the key feature of any programme is the listing of what is being monitored and how that monitoring is to take place and the time-scale over which it should all happen. Typically, and often as an appendix, a monitoring programme will provide a table of locations, dates and sampling methods that are proposed and which, if undertaken in full, will deliver the published monitoring programme.

There are a number of commercial software packages which can assist with the implementation of the programme, monitor its progress and flag up inconsistencies or omissions but none of these can provide the key building block which is the programme itself.

Environmental monitoring data management systems

Given the multiple types and increasing volumes and importance of monitoring data, commercial software Environmental Data Management Systems (EDMS) or E-MDMS are increasingly in common use by regulated industries. They provide a means of managing all monitoring data in a single central place. Quality validation, compliance checking, verifying all data has been received, and sending alerts are generally automated. Typical interrogation functionality enables comparison of data sets both temporarily and spatially. They will also generate regulatory and other reports.

Formal Certification:

(May 2014) there is only one certification scheme specifically for environmental data management software. This is provided by the Environment Agency in the UK under its Monitoring Certification Scheme (MCERTS) see [Environmental certification].

Sampling methods

There are a wide range of sampling methods which depend on the type of environment, the material being sampled and the subsequent analysis of the sample.

At its simplest a sample can be filling a clean bottle with river water and submitting it for conventional chemical analysis. At the more complex end, sample data may be produced by complex electronic sensing devices taking sub-samples over fixed or variable time periods.

Judgmental sampling

In judgmental sampling, the selection of sampling units (i.e., the number and location and/or timing of collecting samples) is based on knowledge of the feature or condition under investigation and on professional judgment. Judgmental sampling is distinguished from probability-based sampling in that inferences are based on professional judgment, not statistical scientific theory. Therefore, conclusions about the target population are limited and depend entirely on the validity and accuracy of professional judgment; probabilistic statements about parameters are not possible. As described in subsequent chapters, expert judgment may also be used in conjunction with other sampling designs to produce effective sampling for defensible decisions.

Simple random sampling

In simple random sampling, particular sampling units (for example, locations and/ or times) are selected using random numbers, and all possible selections of a given number of units are equally likely. For example, a simple random sample of a set of drums can be taken by numbering all the drums and randomly selecting numbers from that list or by sampling an area by using pairs of random coordinates. This method is easy to understand, and the equations for determining sample size are relatively straightforward. Simple random sampling is most useful when the population of interest is relatively homogeneous; i.e., no major patterns of contamination or "hot spots" are expected. The main advantages of this design are:

1. It provides statistically unbiased estimates of the mean, proportions, and variability.
2. It is easy to understand and easy to implement.
3. Sample size calculations and data analysis are very straightforward.

In some cases, implementation of a simple random sample can be more difficult than some other types of designs (for example, grid samples) because of the difficulty of precisely identifying random geographic locations. Additionally, simple random sampling can be more costly than other plans if difficulties in obtaining samples due to location causes an expenditure of extra effort.

Stratified sampling

In stratified sampling, the target population is separated into non-overlapping strata, or subpopulations that are known or thought to be more homogeneous (relative to the environmental medium or the contaminant), so that there tends to be less variation among sampling units in the same stratum than among sampling units in different strata. Strata may be chosen on the basis of spatial or temporal proximity of the units, or on the basis of preexisting information or professional judgment about the site or process. Advantages of this sampling design are that it has potential for achieving greater precision in estimates of the mean and variance, and that it allows computation of reliable estimates for population subgroups of special interest. Greater precision can be obtained if the measurement of interest is strongly correlated with the variable used to make the strata.

Systematic and grid sampling

In systematic and grid sampling, samples are taken at regularly spaced intervals over space or time. An initial location or time is chosen at random, and then the remaining sampling locations are defined so that all locations are at regular intervals over an area (grid) or time (systematic). Examples Systematic Grid Sampling - Square Grid Systematic Grid Sampling - Triangular Grids of systematic grids include square, rectangular, triangular, or radial grids. Cressie, 1993. In random systematic sampling, an initial sampling location (or time) is chosen at random and the remaining sampling sites are specified so that they are located according to a regular pattern. Random systematic sampling is used to search for hot spots and to infer means, percentiles, or other parameters and is also useful for estimating spatial patterns or trends over time. This design provides a practical and easy method for designating sample locations and ensures uniform coverage of a site, unit, or process.

Ranked set sampling is an innovative design that can be highly useful and cost efficient in obtaining better estimates of mean concentration levels in soil and other environmental media by explicitly incorporating the professional judgment of a field investigator or a field screening measurement method to pick specific sampling locations in the field. Ranked set sampling uses a two-phase sampling design that identifies sets of field locations, utilizes inexpensive measurements to rank locations within each set, and then selects one location from each set for sampling. In ranked set sampling, m sets (each of size r) of field locations are identified using simple random sampling. The locations are ranked independently within each set using professional judgment or inexpensive, fast, or surrogate measurements. One sampling unit from each set is then selected (based on the observed ranks) for subsequent measurement using a more accurate and reliable (hence, more expensive) method for the contaminant of interest. Relative to simple random sampling, this design results in more representative samples and so leads to more precise estimates of the population parameters. Ranked set sampling is useful when the cost of locating and ranking locations in the field is low compared to laboratory measurements. It is also appropriate when an inexpensive auxiliary variable (based on expert knowledge or measurement) is available to rank population units with respect to the variable of interest. To use this design effectively, it is important that the ranking method and analytical method are strongly correlated.

Adaptive cluster sampling

In adaptive cluster sampling, samples are taken using simple random sampling, and additional samples are taken at locations where measurements exceed some threshold value. Several additional rounds of sampling and analysis may be needed. Adaptive cluster sampling tracks the selection probabilities for later phases of sampling so that an unbiased estimate of the population mean can be calculated despite oversampling of certain areas. An example application of adaptive cluster sampling is delineating the borders of a plume of contamination. Adaptive sampling is useful for estimating or searching for rare characteristics in a population and is appropriate for inexpensive, rapid measurements. It enables delineating the boundaries of hot spots, while also using all data collected with appropriate weighting to give unbiased estimates of the population mean.

Grab samples

Grab samples are samples taken of a homogeneous material, usually water, in a single vessel. Filling a clean bottle with riverwater is a very common example. Grab samples provide a good snap-shot view of the quality of the sampled environment at the point of sampling and at the time of sampling. Without additional monitoring, the results cannot be extrapolated to other times or to other parts of the river, lake or ground-water.

In order to enable grab samples or rivers to be treated as representative, repeat transverse and longitudinal transect surveys taken at different times of day and times of year are required to establish that the grab-sample location is as representative as is reasonably possible. For large rivers such surveys should also have regard to the depth of the sample and how to best manage the sampling locations at times of flood and drought.

In lakes grab samples are relatively simple to take using depth samplers which can be lowered to a pre-determined depth and then closed trapping a fixed volume of water from the required depth. In all but the shallowest lakes, there are major changes in the chemical composition of lake water at different depths, especially during the summer months when many lakes stratify into a warm, well oxygenated upper layer (epilimnion) and a cool de-oxygenated lower layer (hypolimnion).

In the open seas marine environment grab samples can establish a wide range of base-line parameters such as salinity and a range of cation and anion concentrations. However, where changing conditions are an issue such as near river or sewage discharges, close to the effects of volcanism or close to areas of freshwater input from melting ice, a grab sample can only give a very partial answer when taken on its own.

Semi-continuous monitoring and continuous

There is a wide range of specialized sampling equipment available that can be programmed to take samples at fixed or variable time intervals or in response to an external trigger. For example, a sampler can be programmed to start taking samples of a river at 8 minute intervals when the rainfall intensity rises above 1 mm / hour. The trigger in this case may be a remote rain gauge communicating with the sampler by using cell phone or meteor burst technology. Samplers can also take individual discrete samples at each sampling occasion or bulk up samples into composite so that in the course of one day, such a sampler might produce 12 composite samples each composed of 6 sub-samples taken at 20 minute intervals.

Continuous or quasi-continuous monitoring involves having an automated analytical facility close to the environment being monitored so that results can, if required, be viewed in real time. Such systems are often established to protect important water supplies such as in the River Dee regulation system but may also be part of an overall monitoring strategy on large strategic rivers where early warning of potential problems is essential. Such systems routinely provide data on parameters such as pH, dissolved oxygen, conductivity, turbidity and colour but it is also possible to operate gas liquid chromatography with mass spectrometry technologies (GLC/MS) to examine a wide range of potential organic pollutants. In all examples of automated bank-side analysis there is a requirement for water to be pumped from the river into the monitoring station. Choosing a location for the pump inlet is equally as critical as deciding on the location for a river grab sample. The design of the pump and pipework also requires careful design to avoid artefacts being introduced through the action of pumping the water. Dissolved oxygen concentration is difficult to sustain through a pumped system and GLC/MS facilities can detect micro-organic contaminants from the pipework and glands.

Passive sampling

The use of passive samplers greatly reduces the cost and the need of infrastructure on the sampling location. Passive samplers are semi-disposable and can be produced at a relatively low cost, thus they can be employed in great numbers, allowing for a better cover and more data being collected. Due to being small the passive sampler can also be hidden, and thereby lower the risk of vandalism. Examples of passive sampling devices are the diffusive gradients in thin films (DGT) sampler, Chemcatcher, Polar organic chemical integrative sampler (POCIS), and an air sampling pump.

Remote surveillance

Although on-site data collection using electronic measuring equipment is common-place, many monitoring programmes also use remote surveillance and remote access to data in real time. This requires the on-site monitoring equipment to be connected to a base station via either a telemetry network, land-line, cell phone network or other telemetry system such as Meteor burst. The advantage of remote surveillance is that many data feeds can come into a single base station for storing and analysis. It also enable trigger levels or alert levels to be set for individual monitoring sites and/or parameters so that immediate action can be initiated if a trigger level is exceeded. The use of remote surveillance also allows for the installation of very discrete monitoring equipment which can often be buried, camouflaged or tethered at depth in a lake or river with only a short whip aerial protruding. Use of such equipment tends to reduce vandalism and theft when monitoring in locations easily accessible by the public.

Remote sensing

Environmental remote sensing uses aircraft or satellites to monitor the environment using multi-channel sensors.

There are two kinds of remote sensing. Passive sensors detect natural radiation that is emitted or reflected by the object or surrounding area being observed. Reflected sunlight is the most common source of radiation measured by passive sensors and in environmental remote sensing, the sensors used are tuned to specific wavelengths from far infrared through visible light frequencies to the far ultraviolet. The volumes of data that can be collected are very large and require dedicated computational support . The output of data analysis from remote sensing are false colour images which differentiate small differences in the radiation characteristics of the environment being monitored. With a skilful operator choosing specific channels it is possible to amplify differences which are imperceptible to the human eye. In particular it is possible to discriminate subtle changes in chlorophyll a and chlorophyll b concentrations in plants and show areas of an environment with slightly different nutrient regimes.

Active remote sensing emits energy and uses a passive sensor to detect and measure the radiation that is reflected or backscattered from the target. LIDAR is often used to acquire information about the topography of an area, especially when the area is large and manual surveying would be prohibitively expensive or difficult.

Remote sensing makes it possible to collect data on dangerous or inaccessible areas. Remote sensing applications include monitoring deforestation in areas such as the Amazon Basin, the effects of climate change on glaciers and Arctic and Antarctic regions, and depth sounding of coastal and ocean depths.

Orbital platforms collect and transmit data from different parts of the electromagnetic spectrum, which in conjunction with larger scale aerial or ground-based sensing and analysis, provides information to monitor trends such as El Niño and other natural long and short term phenomena. Other uses include different areas of the earth sciences such as natural resource management, land use planning and conservation.

Bio-monitoring

The use of living organisms as monitoring tools has many advantages. Organisms living in the environment under study are constantly exposed to the physical, biological and chemical influences of that environment. Organisms that have a tendency to accumulate chemical species can often accumulate significant quantities of material from very low concentrations in the environment. Mosses have been used by many investigators to monitor heavy metal concentrations because of their tendency to selectively adsorb heavy metals.

Similarly, eels have been used to study halogenated organic chemicals, as these are adsorbed into the fatty deposits within the eel.

Other sampling methods

Ecological sampling requires careful planning to be representative and as noninvasive as possible. For grasslands and other low growing habitats the use of a quadrat– a 1-metre square frame – is often used with the numbers and types of organisms growing within each quadrat area counted

Sediments and soils require specialist sampling tools to ensure that the material recovered is representative. Such samplers are frequently designed to recover a specified volume of material and may also be designed to recover the sediment or soil living biota as well such as the Ekman grab sampler.

Data interpretations

The interpretation of environmental data produced from a well designed monitoring programme is a large and complex topic addressed by many publications. Regrettably it is sometimes the case that scientists approach the analysis of results with a pre-conceived outcome in mind and use or misuse statistics to demonstrate that their own particular point of view is correct.

Statistics remains a tool that is equally easy to use or to misuse to demonstrate the lessons learnt from environmental monitoring.

Environmental quality indices

Since the start of science-based environmental monitoring, a number of quality indices have been devised to help classify and clarify the meaning of the considerable volumes of data involved. Stating that a river stretch is in "Class B" is likely to be much more informative than stating that this river stretch has a mean BOD of 4.2, a mean dissolved oxygen of 85%, etc. In the UK the Environment Agency formally employed a system called General Quality Assessment (GQA) which classified rivers into six quality letter bands from A to F based on chemical criteria and on biological criteria. The Environment Agency and its devolved partners in Wales (Countryside

Council for Wales, CCW) and Scotland (Scottish Environmental Protection Agency, SEPA) now employ a system of biological, chemical and physical classification for rivers and lakes that corresponds with the EU Water Framework Directive.

GREEN CHEMISTRY

Green chemistry is the design of chemical products and processes that reduce or eliminate the use or generation of hazardous substances. Green chemistry applies across the life cycle of a chemical product, including its design, manufacture, use, and ultimate disposal. Green chemistry is also known as sustainable chemistry.

- Prevents pollution at the molecular level
- Is a philosophy that applies to all areas of chemistry, not a single discipline of chemistry
- Applies innovative scientific solutions to real-world environmental problems
- Results in source reduction because it prevents the generation of pollution
- Reduces the negative impacts of chemical products and processes on human health and the environment
- Lessens and sometimes eliminates hazard from existing products and processes
- Designs chemical products and processes to reduce their intrinsic hazards How green chemistry differs from cleaning up pollution

Green chemistry reduces pollution at its source by minimizing or eliminating the hazards of chemical feedstocks, reagents, solvents, and products.

This is unlike cleaning up pollution (also called remediation), which involves treating waste streams (end-of-the-pipe treatment) or cleanup of environmental spills and other releases. Remediation may include separating hazardous chemicals from other materials, then treating them so they are no longer hazardous or concentrating them for safe disposal. Most remediation activities do not involve green chemistry. Remediation removes hazardous materials from the environment; on the other hand, green chemistry keeps the hazardous materials out of the environment in the first place.

If a technology reduces or eliminates the hazardous chemicals used to clean up environmental contaminants, this technology would qualify as a green chemistry technology. One example is replacing a hazardous sorbent [chemical] used to capture mercury from the air for safe disposal with an effective, but nonhazardous sorbent. Using the nonhazardous sorbent means that the hazardous sorbent is never manufactured and so the remediation technology meets the definition of green chemistry.

Green chemistry's 12 principles

These principles demonstrate the breadth of the concept of green chemistry:

1. Prevent waste: Design chemical syntheses to prevent waste. Leave no waste to treat or clean up.
2. Maximize atom economy: Design syntheses so that the final product contains the maximum proportion of the starting materials. Waste few or no atoms.
3. Design less hazardous chemical syntheses: Design syntheses to use and generate substances with little or no toxicity to either humans or the environment.
4. Design safer chemicals and products: Design chemical products that are fully effective yet have little or no toxicity.
5. Use safer solvents and reaction conditions: Avoid using solvents, separation agents, or other auxiliary chemicals. If you must use these chemicals, use safer ones.
6. Increase energy efficiency: Run chemical reactions at room temperature and pressure whenever possible.
7. Use renewable feedstocks: Use starting materials (also known as feedstocks) that are renewable rather than depletable. The source of renewable feedstocks is often agricultural products or the wastes of other processes; the source of depletable feedstocks is often fossil fuels (petroleum, natural gas, or coal) or mining operations.
8. Avoid chemical derivatives: Avoid using blocking or protecting groups or any temporary modifications if possible. Derivatives use additional reagents and generate waste.
9. Use catalysts, not stoichiometric reagents: Minimize waste by using catalytic reactions. Catalysts are effective in small amounts and can carry out a single reaction many times. They are preferable to stoichiometric reagents, which are used in excess and carry out a reaction only once.
10. Design chemicals and products to degrade after use: Design chemical products to break down to innocuous substances after use so that they do not accumulate in the environment.
11. Analyze in real time to prevent pollution: Include in-process, real-time monitoring and control during syntheses to minimize or eliminate the formation of byproducts.

12. Minimize the potential for accidents: Design chemicals and their physical forms (solid, liquid, or gas) to minimize the potential for chemical accidents including explosions, fires, and releases to the environment.

Green chemistry's roots in the Pollution Prevention Act of 1990

To stop creating pollution in the first place became America's official policy in 1990 with the Federal Pollution Prevention Act .

The law defines source reduction as any practice that:

- Reduces the amount of any hazardous substance, pollutant, or contaminant entering any waste stream or otherwise released into the environment (including fugitive emissions) prior to recycling, treatment, or disposal.
- Reduces the hazards to public health and the environment associated with the release of such substances, pollutants, or contaminants.

The term "source reduction" includes:

- Modifications to equipment or technology
- Modifications to process or procedures
- Modifications, reformulation or redesign of products
- Substitution of raw materials
- Improvements in housekeeping, maintenance, training, or inventory control

Section 2 of the Pollution Prevention Act establishes a pollution prevention hierarchy, saying:

- The Congress hereby declares it to be the national policy of the United States that pollution should be prevented or reduced at the source whenever feasible;
- Pollution that cannot be prevented should be recycled in an environmentally safe manner, whenever feasible;
- Pollution that cannot be prevented or recycled should be treated in an environmentally safe manner whenever feasible; an
- Disposal or other release into the environment should be employed only as a last resort and should be conducted in an environmentally safe manner.

Green chemistry aims to design and produce cost-competitive chemical products and processes that attain the highest level of the pollution-prevention hierarchy by reducing pollution at its source.

For those who are creating and using green chemistry, the hierarchy looks like this:

1. Source Reduction and Prevention of Chemical Hazards
   - Designing chemical products to be less hazardous to human health and the environment*
   - Making chemical products from feedstocks, reagents, and solvents that are less hazardous to human health and the environment*
   - Designing syntheses and other processes with reduced or even no chemical waste
   - Designing syntheses and other processes that use less energy or less water
   - Using feedstocks derived from annually renewable resources or from abundant waste
   - Designing chemical products for reuse or recycling
   - Reusing or recycling chemicals
2. Treating chemicals to render them less hazardous before disposal
3. Disposing of untreated chemicals safely and only if other options are not feasible Chemicals that are less hazardous to human health and the environment are:
   - Less toxic to organisms
   - Less damaging to ecosystems
   - Not persistent or bioaccumulative in organisms or the environment
   - Inherently safer to handle and use because they are not flammable or explosive

Trends

Attempts are being made not only to quantify the greenness of a chemical process but also to factor in other variables such as chemical yield, the price of reaction components, safety in handling chemicals, hardware demands, energy profile and ease of product workup and purification. In one quantitative study, the reduction of nitrobenzene to aniline receives 64 points out of 100 marking it as an acceptable synthesis overall whereas a synthesis of an amide using HMDS is only described as adequate with a combined 32 points.

Green chemistry is increasingly seen as a powerful tool that researchers must use to evaluate the environmental impact of nanotechnology. As nanomaterials are developed, the environmental and human health impacts of both the products themselves and the processes to make them must be considered to ensure their long-term economic viability.

Examples

Green solvents

Solvents are consumed in large quantities in many chemical syntheses as well as for cleaning and degreasing. Traditional solvents are often toxic or are chlorinated. Green solvents, on the other hand, are generally derived from renewable resources and biodegrade to innocuous, often a naturally occurring product.

Synthetic techniques

Novel or enhanced synthetic techniques can often provide improved environmental performance or enable better adherence to the principles of green chemistry. For example, the 2005 Nobel Prize for Chemistry was awarded, to Yves Chauvin, Robert H. Grubbs and Richard R. Schrock, for the development of the metathesis method in organic synthesis, with explicit reference to its contribution to green chemistry and "smarter production." A 2005 review identified three key developments in green chemistry in the field of organic synthesis: use of supercritical carbon dioxide as green solvent, aqueous hydrogen peroxide for clean oxidations and the use of hydrogen in asymmetric synthesis. Some further examples of applied green chemistry are supercritical water oxidation, on water reactions, and dry media reactions.

Bioengineering is also seen as a promising technique for achieving green chemistry goals. A number of important process chemicals can be synthesized in engineered organisms, such as shikimate, a Tamiflu precursor which is fermented by Roche in bacteria. Click chemistry is often cited as a style of chemical synthesis that is consistent with the goals of green chemistry. The concept of 'green pharmacy' has recently been articulated based on similar principles.

Carbon dioxide as blowing agent

In 1996, Dow Chemical won the 1996 Greener Reaction Conditions award for their 100% carbon dioxide blowing agent for polystyrene foam production. Polystyrene foam is a common material used in packing and food transportation. Seven hundred million pounds are produced each year in the United States alone. Traditionally, CFC and other ozone-depleting chemicals were used in the production process of the foam sheets, presenting a serious environmental hazard. Flammable, explosive, and, in some cases toxic hydrocarbons have also been used as CFC replacements, but they present their own problems. Dow Chemical discovered that supercritical carbon dioxide works equally as well as a blowing agent, without the need for hazardous substances, allowing the polystyrene to be more easily recycled.

The COused in the process is reused from other industries, so the net carbon released from the process is zero.

Hydrazine

Addressing principle #2 is the Peroxide Process for producing hydrazine without cogenerating salt. Hydrazine is traditionally produced by the Olin Raschig process from sodium hypochlorite (the active ingredient in many bleaches) and ammonia. The net reaction produces one equivalent of sodium chloride for every equivalent of the targeted product hydrazine:

$$NaOCl + 2\ NH\ ' \rightarrow HN\text{-}NH + NaCl + HO$$

In the greener Peroxide process hydrogen peroxide is employed as the oxidant, the side product being water. The net conversion follows:

$$2\ NH + HO\ ' \rightarrow HN\text{-}NH + 2\ HO$$

Addressing principle #4, this process does not require auxiliary extracting solvents. Methyl ethyl ketone is used as a carrier for the hydrazine, the intermediate ketazide phase separates from the reaction mixture, facilitating workup without the need of an extracting solvent.

1,3-Propanediol

Addressing principle #7 is a green route to 1,3-propanediol, which is traditionally generated from petrochemical precursors. It can be produced from renewable precursors via the bioseparation of 1,3-propanediol using a genetically modified strain of E. coli. This diol is used to make new polyesters for the manufacture of carpets.

Lactide

Lactide

In 2002, Cargill Dow (now NatureWorks) won the Greener Reaction Conditions Award for their improved method for polymerization of polylactic acid. Unfortunately, lactide-base polymers do not perform well and the project was discontinued by Dow soon after the award. Lactic acid is produced by fermenting corn and converted to lactide, the cyclic dimer ester of lactic acid using an efficient, tin-catalyzed cyclization. The L,L-lactide enantiomer is isolated by distillation and polymerized in the melt to make a crystallizable polymer, which has some applications including textiles and apparel, cutlery, and food packaging. Wal-Mart has announced that it is using/will use PLA for its produce packaging. The NatureWorks PLA process substitutes renewable materials for petroleum feedstocks, doesn't require the use of hazardous organic solvents typical in other PLA processes, and results in a high-quality polymer that is recyclable and compostable.

Carpet tile backings

In 2003 Shaw Industries selected a combination of polyolefin resins as the base polymer of choice for EcoWorx due to the low toxicity of its feedstocks, superior adhesion properties, dimensional stability, and its ability to be recycled. The EcoWorx compound also had to be designed to be compatible with nylon carpet fiber. Although EcoWorx may be recovered from any fiber type, nylon-6 provides a significant advantage. Polyolefins are compatible with known nylon-6 depolymerization methods. PVC interferes with those processes. Nylon-6 chemistry is well-known and not addressed in first-generation production. From its inception, EcoWorx met all of the design criteria necessary to satisfy the needs of the marketplace from a performance, health, and environmental standpoint. Research indicated that separation of the fiber and backing through elutriation, grinding, and air separation proved to be the best way to recover the face and backing components, but an infrastructure for returning postconsumer EcoWorx to the elutriation process was necessary. Research also indicated that the postconsumer carpet tile had a positive economic value at the end of its useful life. EcoWorx is recognized by MBDC as a certified cradle-to-cradle design.

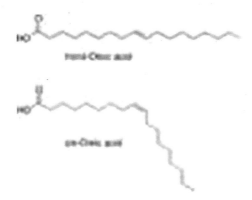

Trans and cis fatty acids

Transesterification of fats

In 2005, Archer Daniels Midland (ADM) and Novozymes won the Greener Synthetic Pathways Award for their enzymeinteresterification process. In response to the U.S. Food and Drug Administration (FDA) mandated labeling of trans-fats on nutritional information by January 1, 2006, Novozymes and ADM worked together to develop a clean, enzymatic process for the interesterification of oils and fats by interchanging saturated and unsaturated fatty acids. The result is commercially variable products without trans-fats. In addition to the human health benefits of eliminating trans-fats, the process has reduced the use of toxic chemicals and water, prevents vast amounts of byproducts, and reduces the amount of fats and oils wasted.

Bio-succinic acid

In 2011, the Outstanding Green Chemistry Accomplishments by a Small Business Award went to BioAmber Inc. for integrated production and downstream applications of bio-based succinic acid. Succinic acid is a platform chemical that is an important starting material in the formulations of everyday products. Traditionally, succinic acid is produced from petroleum-based feedstocks. BioAmber has developed process and technology that produces succinic acid from the fermentation of renewable feedstocks at a lower cost and lower energy expenditure than the petroleum equivalent while sequestering CO rather than emitting it.

Laboratory chemicals

Several laboratory chemicals are controversial from the perspective of Green chemistry. The Massachusetts Institute of Technology created a "Green" Alternatives Wizard [2] to help identify alternatives. Ethidium bromide, xylene, mercury, and formaldehyde have been identified as "worst offenders" which have alternatives.Solvents in particular make a large contribution to the environmental impact of chemical manufacturing and there is a growing focus on introducing Greener solvents into the earliest stage of development of these processes: laboratory-scale reaction and purification methods. In the Pharmaceutical Industry, both GSK and Pfizer have published Solvent Selection Guides for their Drug Discovery chemists.

LEGISLATION

In 2007, The EU put into place the Registration, Evaluation, Authorisation, and Restriction of Chemicals (REACH) program, which requires companies to provide data showing that their products are safe. This regulation (1907/2006) ensures not only the assessment of the chemicals' hazards as well as risks during their uses but also includes measures for banning or restricting/ authorising uses of specific substances. ECHA, the EU Chemicals Agency in Helsinki, is implementing the regulation whereas the enforcement lies with the EU member states.

United States

The U.S. law that governs the majority of industrial chemicals (excluding pesticides, foods, and pharmaceuticals) is the Toxic Substances Control Act (TSCA) of 1976. Examining the role of regulatory programs in shaping the development of green chemistry in the United States, analysts have revealed structural flaws and long-standing weaknesses in TSCA; for example, a 2006 report to the California Legislature concludes that TSCA has produced a domestic chemicals market that discounts the hazardous properties of chemicals relative to their function, price, and performance. Scholars have argued that such market conditions represent a key barrier to the scientific, technical, and commercial success of green chemistry in the U.S., and fundamental policy changes are needed to correct these weaknesses.

Passed in 1990, the Pollution Prevention Act helped foster new approaches for dealing with pollution by preventing environmental problems before they happen.

In 2008, the State of California approved two laws aiming to encourage green chemistry, launching the California Green Chemistry Initiative. One of these statutes required California's Department of Toxic Substances Control (DTSC) to develop new regulations to prioritize "chemicals of concern" and promote the substitution of hazardous chemicals with safer alternatives. The resulting regulations took effect in 2013, initiating DTSC's Safer Consumer Products Program.

3

Environmental Health

Environmental health is the branch of public health that is concerned with all aspects of the natural and built environment that may affect human health. Health is the science, practice, and study of a human's well-being and their health and preventing illnesses and human injuries. Other terms referring to or concerning environmental health are environmental public health, and public health protection / environmental health protection. Environmental health and environmental protection are very much related. Environmental health is focused on the natural and built environments for the benefit of human health, whereas environmental protection is concerned with protecting the natural environment for the benefit of human health and the ecosystem. Research in the environmental health field tries to limit the harmful exposures through natural things such as soil, water, air food, etc. The definition of environmental health varies from organization to organization, although the basic premise remains the same. Below are definitions from various federal and nonfederal organizations/agencies.

Environmental health addresses all the physical, chemical, and biological factors external to a person, and all the related factors impacting behaviors. It encompasses the assessment and control of those environmental factors that can potentially affect health. It is targeted towards preventing disease and creating health-supportive environments. This definition excludes behavior not related to environment, as well as behavior related to the social and cultural environment, and genetics. – World Health Organization

Environmental health and protection refers to protection against environmental factors that may adversely impact human health or the ecological balances essential to long-term human health and environmental quality, whether in the natural or man-made environment. – National Environmental Health Association

Environmental Health is the field of science that studies how the environment influences human health and disease. "Environment," in this context, means things in the natural environment like air, water and soil, and also all the physical, chemical, biological and social features of our surroundings.

The man-made, or "built," environment includes physical structures where people live and work such as homes, offices, schools, farms and factories, as well as community systems such as roads and transportation systems, land use practices and waste management. Consequences of human alteration to the natural environment, such as air pollution, are also parts of the man-made environment.

The social environment encompasses lifestyle factors like diet and exercise, socioeconomic status, and other societal influences that may affect health. – National Institute of Environmental Health Science

ENVIRONMENTAL HEALTH PROFESSION

Environmental health professionals may be known as environmental health officers, public health inspectors, environmental health specialists, environmental health practitioners, or sanitarians. Researchers and policy-makers also play important roles in how environmental health is practiced in the field. In many European countries, physicians and veterinarians are involved in environmental health. In the United Kingdom, practitioners must have a graduate degree in environmental health and be certified and registered with the Chartered Institute of Environmental Health or the Royal Environmental Health Institute of Scotland. In Canada, practitioners in environmental health are required to obtain an approved bachelor's degree in environmental health along with the national professional certificate, the Certificate in Public Health Inspection (Canada) CPHI(C). Many states in the United States also require that individuals have a bachelor's degree and professional licenses in order to practice environmental health. California state law defines the scope of practice of environmental health as follows:

"Scope of practice in environmental health" means the practice of environmental health by registered environmental health specialists in the public and private sector within the meaning of this article and includes, but is not limited to, organization, management, education, enforcement, consultation, and emergency response for the purpose of prevention of environmental health hazards and the promotion and protection of the public health and the environment in the following areas: food protection; housing; institutional environmental health; land use; community noise control; recreational swimming areas and waters; electromagnetic radiation control; solid, liquid, and hazardous materials management; underground storage tank control; onsite septic systems; vector control; drinking water quality; water sanitation; emergency preparedness; and milk and dairy sanitation pursuant to Section 33113 of the Food and Agricultural Code.

The environmental health profession had its modern-day roots in the sanitary and public health movement of the United Kingdom. This was epitomized by Sir Edwin Chadwick, who was instrumental in the repeal of the poor laws, and in 1884 was the founding president of the Association of Public Sanitary Inspectors, now called the Chartered Institute of Environmental Health.

ENVIRONMENTAL EPIDEMIOLOGY

Environmental epidemiology is a branch of epidemiology concerned with the discovery of the environmental exposures that contribute to or protect against injuries, illnesses, developmental conditions, disabilities, and deaths; and identification of public health and health care actions to manage the risks associated with harmful exposures.

Environmental epidemiology studies external factors that affect the incidence, prevalence, and geographic range of health conditions. These factors may be naturally occurring or may be introduced into environments where people live, work, and play. Environmental exposures are involuntary and thus generally exclude occupational exposures (covered by occupational epidemiology) and voluntary exposures such as active smoking, medications, and diet.

Environmental exposures can be broadly categorized into those that are proximate (e.g., directly leading to a health condition), including chemicals, physical agents, and microbiological pathogens, and those that are distal, such as socioeconomic conditions, climate change, and other broad-scale environmental changes. Proximate exposures occur through air, food, water, and skin contact. Distal exposures cause adverse health conditions directly by altering proximate exposures, and indirectly through changes in ecosystems and other support systems for human health.

Environmental epidemiology research can inform risk assessments; development of standards and other risk management activities; and estimates of the co-benefits and co-harms of policies designed to reduce global environment change, including policies implemented in other sectors (e.g. food and water) that can affect human health.

Vulnerability is the summation of all risk and protective factors that ultimately determine whether an individual or subpopulation experiences adverse health outcomes when an exposure to an environmental agent occurs. Sensitivity is an individual's or subpopulation's increased responsiveness, primarily for biological reasons, to that exposure. Biological sensitivity may be related to developmental stage, pre-existing medical conditions, acquired factors, and genetic factors. Socioeconomic factors also play a critical role in altering vulnerability and sensitivity to environmentally mediated factors by increasing the likelihood of exposure to harmful agents, interacting with biological factors that mediate risk, and/or leading to differences in the ability to prepare for or cope with exposures or early phases of illness. Populations living in certain regions may be at increased risk due to location and the environmental characteristics of a region.

Toxicology

Toxicology is a discipline, overlapping with biology, chemistry, pharmacology, and medicine, that involves the study of the adverse effects of chemical substances on living organisms and the practice of diagnosing and treating exposures to toxins and toxicants. The relationship between dose and its effects on the exposed organism is of high significance in toxicology. Factors that influence chemical toxicity include the dosage (and whether it is acute or chronic), route of exposure, species, age, sex, and environment. Toxicologists are experts on poisons and poisoning.

BASIC PRINCIPLES

The goal of toxicity assessment is to identify adverse effects of a substance. Adverse effects depend on two main factors: i) routes of exposure (oral, inhalation, or dermal) and ii) dose (duration and concentration of exposure). To explore dose, substances are tested in both acute and chronic models. Generally, different sets of experiments are conducted to determine whether a substance causes cancer and to examine other forms of toxicity.

Factors that influence chemical toxicity:

- Dosage
- Both large single exposures (acute) and continuous small exposures (chronic) are studied.
- Route of exposure
- Ingestion, inhalation or skin absorption
- Other factors
- Species
- Age
- Sex
- Health
- Environment
- Individual characteristics

Testing methods

Toxicity experiments may be conducted in vivo (using the whole animal) or in vitro (testing on isolated cells or tissues), or in silico (in a computer simulation).

Non-human animals

The classic experimental tool of toxicology is testing on non-human animals. An example of a model organism is Galleria mellonella, which can replace small mammals to study toxicology in vivo. As of 2014, such animal testing provides information that is not available by other means about how substances function in a living organism.

Alternative testing methods

While testing in animal models remains as a method of estimating human effects, there are both ethical and technical concerns with animal testing.

Since the late 1950s, the field of toxicology has sought to reduce or eliminate animal testing under the rubric of "Three Rs" - reduce the number of experiments with animals to the minimum necessary; refine experiments to cause less suffering, and replace in vivo experiments with other types, or use more simple forms of life when possible.

Computer modeling is an example of alternative testing methods; using computer models of chemicals and proteins, structure-activity relationships can be determined, and chemical structures that are likely to bind to, and interfere with, proteins with essential functions, can be identified. This work requires expert knowledge in molecular modeling and statistics together with expert judgment in chemistry, biology and toxicology.

In 2007 the National Academy of Sciences published a report called "T oxicity Testing in the 21st Century: A Vision and a Strategy" which opened with a statement: "Change often involves a pivotal event that builds on previous history and opens the door to a new era. Pivotal events in science include the discovery of penicillin, the elucidation of the DNA double helix, and the development of computers. ...Toxicity testing is approaching such a scientific pivot point. It is poised to take advantage of the revolutions in biology and biotechnology. Advances in toxicogenomics, bioinformatics, systems biology, epigenetics, and computational toxicology could transform toxicity testing from a system based on whole-animal testing to one founded primarily on in vitro methods that evaluate changes in biologic processes using cells, cell lines, or cellular components, preferably of human origin." As of 2010 that vision was still unrealized. As of 2014 that vision was still unrealized.

In some cases shifts away from animal studies has been mandated by law or regulation; the European Union (EU) prohibited use of animal testing for cosmetics in 2013.

Dose response complexities

Most chemicals display a classic dose response curve – at a low dose (below a threshold), no effect is observed. Some show a phenomenon known as sufficient challenge – a small exposure produces animals that "grow more rapidly , have better general appearance and coat quality, have fewer tumors, and live longer than the control animals". A few chemicals have no well-defined safe level of exposure. These are treated with special care. Some chemicals are subject to bioaccumulation as they are stored in rather than being excreted from the body; these also receive special consideration.

Several measures are commonly used to describe toxic dosages according to the degree of effect on an organism or a population, and some are specifically defined by various laws or organizational usage. These include:

- LD50 = Median lethal dose, a dose that will kill 50% of an exposed population
- NOEL = No Observed Effect Level, the highest dose known to show no effect
- NOAEL = No Observed Adverse Effect Level, the highest dose known to show no adverse effects
- PEL = Permissable Exposure Limit, the highest concentration permitted under US OSHA regulations
- STEL = Short-Term Exposure Limit, the highest concentration permitted for short periods of time, in general 15–30 minutes
- TWA = Time-Weighted Average, the average amount of an agent's concentration over a specified period of time, usually 8 hours.
- TTC = Threshold of Toxicological Concern have been established for the constituents of tobacco smoke

Types

Medical toxicology

Medical toxicology is the discipline that requires physician status (MD or DO degree plus specialty education and experience).

Clinical toxicology

"Clinical toxicology" redirects here. For the journal, see Clinical Toxicology. Clinical toxicology is the discipline that can be practiced not only by physicians but also other health professionals with a master's degree in clinical toxicology: physician extenders (physician assistants, nurse practitioners), nurses, pharmacists, and allied health professionals.

Computational toxicology

Computational toxicology is a discipline that develops mathematical and computer-based models to better understand and predict adverse health effects caused by chemicals, such as environmental pollutants and pharmaceuticals. Within the Toxicology in the 21st Century project, the best predictive models were identified to be Deep Neural Networks, Random Forest, and Support Vector Machines, which can reach the performance of in vitro experiments.

Toxicology as a profession

A toxicologist is a scientist or medical personnel who specializes in the study of symptoms, mechanisms, treatments and detection of venoms and toxins; especially the poisoning of people. To work as a toxicologist one should obtain a degree in toxicology or a related degree like biology, chemistry, pharmacology or biochemistry. Toxicologists perform many different duties including research in the academic, nonprofit and industrial fields, product safety evaluation, consulting, public service and legal regulation.

Requirements

To work as a toxicologist one should obtain a degree in toxicology or a related degree like biology, chemistry or biochemistry. Bachelor's degree programs in toxicology cover the chemical makeup of toxins and their effects on biochemistry, physiology and ecology. After introductory life science courses are complete, students typically enroll in labs and apply toxicology principles to research and other studies. Advanced students delve into specific sectors, like the pharmaceutical industry or law enforcement, which apply methods of toxicology in their work. The Society of Toxicology (SOT) recommends that undergraduates in postsecondary schools that don't offer a bachelor's degree in toxicology consider attaining a degree in biology or chemistry. Additionally, the SOT advises aspiring toxicologists to take statistics and mathematics courses, as well as gain laboratory experience through lab courses, student research projects and internships.

Duties

Toxicologists perform many duties including research in the academic, nonprofit and industrial fields; product safety evaluation, consulting, public service and legal regulation. In order to research and assess the effects of chemicals, toxicologists perform carefully designed studies and experiments.

These experiments help identify the specific amount of a chemical that may cause harm and potential risks of being near or using products that contain certain chemicals. Research projects may range from assessing the effects of toxic pollutants on the environment to evaluating how the human immune system responds to chemical compounds within pharmaceutical drugs. While the basic duties of toxicologists are to determine the effects of chemicals on organisms and their surroundings, specific job duties may vary based on industry and employment. For example, forensic toxicologists may look for toxic substances in a crime scene, whereas aquatic toxicologists may analyze the toxicity level of water bodies.

Compensation

The salary for jobs in toxicology is dependent on several factors, including level of schooling, specialization, experience. The U.S. Bureau of Labor Statistics (BLS) notes that jobs for biological scientists, which generally include toxicologists, were expected to increase by 21% between 2008 and 2018. The BLS notes that this increase could be due to research and development growth in biotechnology, as well as budget increases for basic and medical research in biological science.

Etymology and pronunciation

The word toxicology (/ÌtRksjÈkRlYd′i/) is a neoclassical compound from New Latin, first attested circa 1799, from the combining forms toxico- + -logy, which in turn come from the Ancient Greek words ôïñéêüò toxikos, "poisonous", and ëüãïò logos, "subject matter").

Exposure science

Exposure science is the study of an organism's (usually human) contact with chemical, physical, biological agents or other health risk (eg accidental) occurring in their environments, and advances knowledge of the mechanisms and dynamics of events either causing or preventing adverse health outcomes.

Exposure science plays a fundamental role in the development and application of epidemiology, toxicology, and risk assessment. It provides critical information for protecting human and ecosystem health. Exposure science also has the ability to play an effective role in other fields, including environmental regulation, urban, traffic safety and ecosystem planning, and disaster management; in many cases these are untapped opportunities. Exposure science links human and ecologic behavior to environmental processes in such a way that the information generated can be used to mitigate or prevent future adverse exposures.

Environmental engineering

Environmental engineering is the branch of engineering concerned with the application of scientific and engineering principles for protection of human populations from the effects of adverse environmental factors; protection of environments, both local and global, from potentially deleterious effects of natural and human activities; and improvement of environmental quality.

Environmental engineering can also be described as a branch of applied science and technology that addresses the issues of energy preservation, protection of assets and control of waste from human and animal activities. Furthermore, it is concerned with finding plausible solutions in the field of public health, such as waterborne diseases, implementing laws which promote adequate sanitation in urban, rural and recreational areas. It involves waste water management, air pollution control, recycling, waste disposal, radiation protection, industrial hygiene, animal agriculture, environmental sustainability, public health and environmental engineering law. It also includes studies on the environmental impact of proposed construction projects.

Environmental engineers study the effect of technological advances on the environment. To do so, they conduct studies on hazardous-waste management to evaluate the significance of such hazards, advise on treatment and containment, and develop regulations to prevent mishaps. Environmental engineers design municipal water supply and industrial wastewater treatment systems. They address local and worldwide environmental issues such as the effects of acid rain, global warming, ozone depletion, water pollution and air pollution from automobile exhausts and industrial sources.

Many universities offer environmental engineering programs at either the department of civil engineering or the department of chemical engineering at engineering faculties. Environmental "civil" engineers focus on hydrology , water resources management, bioremediation, and water treatment plant design. Environmental "chemical" engineers, on the other hand, focus on environmental chemistry, advanced air and water treatment technologies and separation processes.

More engineers are obtaining specialized training in law (J.D.) and are utilizing their technical expertise in the practices of environmental engineering law.

Most jurisdictions also impose licensing and registration requirements.

Development

Ever since people first recognized that their health is related to the quality of their environment, they have applied principles to attempt to improve the quality of their environment. The ancient Indian Harappan civilization utilized early sewers in some cities more than 5000 years ago. The Romans constructed aqueducts to prevent drought and to create a clean, healthful water supply for the metropolis of Rome. In the 15th century, Bavaria created laws restricting the development and degradation of alpine country that constituted the region's water supply.

The field emerged as a separate environmental discipline during the middle third of the 20th century in response to widespread public concern about water and pollutant and increasingly extensive environmental quality degradation. However, its roots extend back to early efforts in public health engineering. Modern environmental engineering began in London in the mid-19th century when Joseph Bazalgette designed the first major sewerage system that reduced the incidence of waterborne diseases such as cholera. The introduction of drinking water treatment and sewage treatment in industrialized countries reduced waterborne diseases from leading causes of death to rarities.

In many cases, as societies grew, actions that were intended to achieve benefits for those societies had longer-term impacts which reduced other environmental qualities. One example is the widespread application of the pesticide DDT to control agricultural pests in the years following World War II. While the agricultural benefits were outstanding and crop yields increased dramatically thus reducing world hunger substantially, and malaria was controlled better than it ever had been, numerous species were brought to the verge of extinction due to the impact of the DDT on their reproductive cycles. The story of DDT as vividly told in Rachel Carson's Silent Spring (1962) is considered to be the birth of the modern environmental movement and of the modern field of "environmental engineering."

Conservation movements and laws restricting public actions that would harm the environment have been developed by various societies for millennia. Notable examples are the laws decreeing the construction of sewers in London and Paris in the 19th century and the creation of the U.S. national park system in the early 20th century.

SCOPE

Solid waste management

Environmental impact assessment and mitigation

Scientists have air pollution dispersion models to evaluate the concentration of a pollutant at a receptor or the impact on overall air quality from vehicle exhausts and industrial flue gas stack emissions. To some extent, this field overlaps the desire to decrease carbon dioxide and other greenhouse gas emissions from combustion processes. They apply scientific and engineering principles to evaluate if there are likely to be any adverse impacts to water quality, air quality, habitat quality, flora and fauna, agricultural capacity, traffic impacts, social impacts, ecological impacts, noise impacts, visual (landscape) impacts, etc. If impacts are expected, they then develop mitigation measures to limit or prevent such impacts. An example of a mitigation measure would be the creation of wetlands in a nearby location to mitigate the filling in of wetlands necessary for a road development if it is not possible to reroute the road.

In the United States, the practice of environmental assessment was formally initiated on January 1, 1970, the effective date of the National Environmental Policy Act(NEPA). Since that time, more than 100 developing and developed nations either have planned specific analogous laws or have adopted procedure used elsewhere. NEPA is applicable to all federal agencies in the United States.

Water supply and treatment

Sewage treatment plant, Australia

Engineers evaluate the water balance within a watershed and determine the available water supply, the water needed for various needs in that watershed, the seasonal cycles of water movement through the watershed and they develop systems to store, treat, and convey water for various uses. Water is treated to achieve water quality objectives for the end uses. In the case of a potable water supply, water is treated to minimize the risk of infectious disease transmission, the risk of non-infectious illness, and to create a palatable water flavor. Water distribution systems are designed and built to provide adequate water pressure and flow rates to meet various end-user needs such as domestic use, fire suppression, and irrigation.

Wastewater treatment

Water pollution

There are numerous wastewater treatment technologies. A wastewater treatment train can consist of a primary clarifier system to remove solid and floating materials, a secondary treatment system consisting of an aeration basin followed by flocculation and sedimentation or an activated sludge system and a secondary clarifier, a tertiary biological nitrogen removal system, and a final disinfection process. The aeration basin/ activated sludge system removes organic material by growing bacteria (activated sludge). The secondary clarifier removes the activated sludge from the water. The tertiary system, although not always included due to costs, is becoming more prevalent to remove nitrogen and phosphorus and to disinfect the water before discharge to a surface water stream or ocean outfall.

Air pollution management

Scientists have developed air pollution dispersion models to evaluate the concentration of a pollutant at a receptor or the impact on overall air quality from vehicle exhausts and industrial flue gas stack emissions. To some extent, this field overlaps the desire to decrease carbon dioxide and other greenhouse gas emissions from combustion processes.

Environmental Protection Agency

The U.S. Environmental Protection Agency (EPA) is one of the many agencies that work with environmental engineers to solve key issues. An important component of EPA's mission is to protect and improve air, water, and overall environmental quality in order to avoid or mitigate the consequences of harmful effects.

Ecological engineering for sustainable agriculture in arid and semiarid West African regions

Ecological engineering offers new alternatives for the management of agricultural systems that are more tailored to the ever-changing social and environmental necessities in these regions. This requires managing the complexity of agrosystems, while striving to mimic the functioning of natural ecosystems of West African drylands and taking advantage of traditional practices and local know-how resulting from a long process of adaptation to environmental constraints.

1. Acting on biodiversity. Biodiversity is essential to the productivity of ecosystems and their temporal stability under the impact of external disturbances. Several ecological processes related to biodiversity may be intensified for the benefit of agrosilvopastoral systems: promoting diversity and soil microorganism activity to benefit plants, associating and utilizing the mutual benefits of plants

2. Utilizing organic matter and nutrient cycles. The productivity of agrosystems with low chemical input use in dryland regions is primarily based on efficient organic resource management, and in turn on the nutrient and energy flows they induce. It is thus possible to intervene at several levels: enhancing crop-livestock farming integration to preserve natural resources, restoring the biological activity of soils via specific organic inputs, supplying nutrients to plants locally.

3. Enhancing available water use. Water supplies are limited and irregular in dryland areas. Current management of these supplies—which involves capturing rainwater and surface runoff—could be improved in several ways: adapting to erratic rainfall or drought risks by focusing on: (i) the organization of the farm and community (farm plot patterns in association with the random rainfall distribution, etc.), and on (ii) cropping techniques to reduce crop water needs (plant choices, weeding, etc.), preserving water in crop fields by hampering runoff, accounting for the essential role of trees regarding soil and water in drylands.

4. Managing landscapes and associated ecological processes. Ecological crop pest regulation by their natural enemies is one ecosystem service provided by biodiversity. Better pest management could be considered in association with promoting biodiversity at different scales, e.g. from the plant to the landscape.

EDUCATION

Courses aimed at developing graduates with specific skills in environmental systems or environmental technology are becoming more common and fall into broad classes:

- Mechanical engineering courses oriented towards designing machines and mechanical systems for environmental use such as water treatment facilities, pumping stations, garbage segregation plants and other mechanical facilities;
- Environmental engineering or environmental systems courses oriented towards a civil engineering approach in which structures and the landscape are constructed to blend with or protect the environment;
- Environmental chemistry, sustainable chemistry or environmental chemical engineering courses oriented towards understanding the effects (good and bad) of chemicals in the environment. Focus on mining processes, pollutants and commonly also cover biochemical processes;
- Environmental technology courses oriented towards producing electronic or electrical graduates capable of developing devices and artifacts able to monitor, measure, model and control environmental impact, including monitoring and managing energy generation from renewable sources.

Environmental law

Environmental law, also known as environmental and natural resources law, is a collective term describing the network of treaties, statutes, regulations, common and customary laws addressing the effects of human activity on the natural environment. The core environmental law regimes address environmental pollution. A related but distinct set of regulatory regimes, now strongly influenced by environmental legal principles, focus on the management of specific natural resources, such as forests, minerals, or fisheries. Other areas, such as environmental impact assessment, may not fit neatly into either category, but are nonetheless important components of environmental law.

POLLUTION CONTROL

Air quality

Air quality laws govern the emission of air pollutants into the atmosphere. A specialized subset of air quality laws regulate the quality of air inside buildings. Air quality laws are often designed specifically to protect human health by limiting or eliminating airborne pollutant concentrations. Other initiatives are designed to address broader ecological problems, such as limitations on chemicals that affect the ozone layer, and emissions trading programs to address acid rain or climate change. Regulatory efforts include identifying and categorizing air pollutants, setting limits on acceptable emissions levels, and dictating necessary or appropriate mitigation technologies.

Water quality

Water quality laws govern the release of pollutants into water resources, including surface water, ground water, and stored drinking water. Some water quality laws, such as drinking water regulations, may be designed solely with reference to human health. Many others, including restrictions on the alteration of the chemical, physical, radiological, and biological characteristics of water resources, may also reflect efforts to protect aquatic ecosystems more broadly. Regulatory efforts may include identifying and categorizing water pollutants, dictating acceptable pollutant concentrations in water resources, and limiting pollutant discharges from effluent sources. Regulatory areas include sewage treatment and disposal, industrial and agriculturalwaste water management, and control of surface runoff from construction sites and urban environments.

Waste management

Waste management laws govern the transport, treatment, storage, and disposal of all manner of waste, including municipal solid waste, hazardous waste, and nuclear waste, among many other types. Waste laws are generally designed to minimize or eliminate the uncontrolled dispersal of waste materials into the environment in a manner that may cause ecological or biological harm, and include laws designed to reduce the generation of waste and promote or mandate waste recycling. Regulatory efforts include identifying and categorizing waste types and mandating transport, treatment, storage, and disposal practices.

Contaminant cleanup

Environmental cleanup laws govern the removal of pollution or contaminants from environmental media such as soil, sediment, surface water, or ground water. Unlike pollution control laws, cleanup laws are designed to respond after-the-fact to environmental contamination, and consequently must often define not only the necessary response actions, but also the parties who may be responsible for undertaking (or paying for) such actions. Regulatory requirements may include rules for emergency response, liability allocation, site assessment, remedial investigation, feasibility studies, remedial action, post-remedial monitoring, and site reuse.

Chemical safety

Chemical safety laws govern the use of chemicals in human activities, particularly man-made chemicals in modern industrial applications. As contrasted with media-oriented environmental laws (e.g., air or water quality laws), chemical control laws seek to manage the (potential) pollutants themselves. Regulatory efforts include banning specific chemical constituents in consumer products (e.g., Bisphenol A in plastic bottles), and regulating pesticides.

RESOURCE SUSTAINABILITY

Impact assessment

Environmental impact assessment (EA) is the assessment of the environmental consequences (positive and negative) of a plan, policy, program, or actual projects prior to the decision to move forward with the proposed action. In this context, the term "environmental impact assessment" (EIA) is usually used when applied to actual projects by individuals or companies and the term "strategic environmental assessment" (SEA) applies to policies, plans and programmes most often proposed by organs of state. Environmental assessments may be governed by rules of administrative procedure regarding public participation and documentation of decision making, and may be subject to judicial review.

Water resources

Water resources laws govern the ownership and use of water resources, including surface water and ground water. Regulatory areas may include water conservation, use restrictions, and ownership regimes.

Mineral resources

Mineral resource laws cover several basic topics, including the ownership of the mineral resource and who can work them. Mining is also affected by various regulations regarding the health and safety of miners, as well as the environmental impact of mining.

Forest resources

Forestry laws govern activities in designated forest lands, most commonly with respect to forest management and timber harvesting. Ancillary laws may regulate forest land acquisition and prescribed burn practices. Forest management laws generally adopt management policies, such as multiple use and sustained yield, by which public forest resources are to be managed. Governmental agencies are generally responsible for planning and implementing forestry laws on public forest lands, and may be involved in forest inventory, planning, and conservation, and oversight of timber sales. Broader initiatives may seek to slow or reverse deforestation.

Wildlife and plants

Wildlife laws govern the potential impact of human activity on wild animals, whether directly on individuals or populations, or indirectly via habitat degradation. Similar laws may operate to protect plant species. Such laws may be enacted entirely to protect biodiversity, or as a means for protecting species deemed important for other reasons. Regulatory efforts may including the creation of special conservation statuses, prohibitions on killing, harming, or disturbing protected species, efforts to induce and support species recovery, establishment of wildlife refuges to support conservation, and prohibitions on trafficking in species or animal parts to combat poaching.

Fish and game

Fish and game laws regulate the right to pursue and take or kill certain kinds of fish and wild animal (game). Such laws may restrict the days to harvest fish or game, the number of animals caught per person, the species harvested, or the weapons or fishing gear used. Such laws may seek to balance dueling needs for preservation and harvest and to manage both environment and populations of fish and game. Game laws can provide a legal structure to collect license fees and other money which is used to fund conservation efforts as well as to obtain harvest information used in wildlife management practice.

Principles

Environmental law has developed in response to emerging awareness of and concern over issues impacting the entire world. While laws have developed piecemeal and for a variety of reasons, some effort has gone into identifying key concepts and guiding principles common to environmental law as a whole. The principles discussed below are not an exhaustive list and are not universally recognized or accepted. Nonetheless, they represent important principles for the understanding of environmental law around the world.

Sustainable development

Defined by the United Nations Environment Programme as "development that meets the needs of the present without compromising the ability of future generations to meet their own needs," sustainable development may be considered together with the concepts of "integration" (development cannot be considered in isolation from sustainability) and "interdependence" (social and economic development, and environmental protection, are interdependent). Laws mandating environmental impact assessment and requiring or encouraging development to minimize environmental impacts may be assessed against this principle.

The modern concept of sustainable development was a topic of discussion at the 1972 United Nations Conference on the Human Environment (Stockholm Conference), and the driving force behind the 1983 World Commission on Environment and Development (WCED, or Bruntland Commission). In 1992, the first UN Earth Summit resulted in the Rio Declaration, Principle 3 of which reads: "The right to development must be fulfilled so as to equitably meet developmental and environmental needs of present and future generations." Sustainable development has been a core concept of international environmental discussion ever since, including at the World Summit on Sustainable Development (Earth Summit 2002), and the United Nations Conference on Sustainable Development (Earth Summit 2012, or Rio+20).

Equity

Defined by UNEP to include intergenerational equity - "the right of future generations to enjoy a fair level of the common patrimony" - and intragenerational equity - "the right of all people within the current generation to fair access to the current generation's entitlement to the Earth's natural resources" - environmental equity considers the present generation under an obligation to account for long-term impacts of activities, and to act to sustain the global environment and resource base for future generations. Pollution control and resource management laws may be assessed against this principle.

Transboundary responsibility

Defined in the international law context as an obligation to protect one's own environment, and to prevent damage to neighboring environments, UNEP considers transboundary responsibility at the international level as a potential limitation on the rights of the sovereign state. Laws that act to limit externalities imposed upon human health and the environment may be assessed against this principle.

Public participation and transparency

Identified as essential conditions for "accountable governments,... industrial concerns," and or ganizations generally, public participation and transparency are presented by UNEP as requiring "effective protection of the human right to hold and express opinions and to seek, receive and impart ideas,... a right of access to appropriate, comprehensible and timely information held by governments and industrial concerns on economic and social policies regarding the sustainable use of natural resources and the protection of the environment, without imposing undue financial burdens upon the pplicants and with adequate protection of privacy and business confidentiality," and "effective judicial and administrative proceedings." These principles are present in environmental impact assessment, laws requiring publication and access to relevant environmental data, and administrative procedure.

Precautionary principle

One of the most commonly encountered and controversial principles of environmental law, the Rio Declaration formulated the precautionary principle as follows:

In order to protect the environment, the precautionary approach shall be widely applied by States according to their capabilities. Where there are threats of serious or irreversible damage, lack of full scientific certainty shall not be used as a reason for postponing cost-effective measures to prevent environmental degradation.

The principle may play a role in any debate over the need for environmental regulation.

Prevention

The concept of prevention . . . can perhaps better be considered an overarching aim that gives rise to a multitude of legal mechanisms, including prior assessment of environmental harm, licensing or authorization that set out the conditions for operation and the consequences for violation of the conditions, as well as the adoption of strategies and policies. Emission limits and other product or process standards, the use of best available techniques and similar techniques can all be seen as applications of the concept of prevention.

Polluter pays principle

The polluter pays principle stands for the idea that "the environmental costs of economic activities, including the cost of preventing potential harm, should be internalized rather than imposed upon society at large." All issues related to responsibility for cost for environmental remediation and compliance with pollution control regulations involve this principle.

Theory

Environmental law is a continuing source of controversy. Debates over the necessity, fairness, and cost of environmental regulation are ongoing, as well as regarding the appropriateness of regulations vs. market solutions to achieve even agreed-upon ends.

Allegations of scientific uncertainty fuel the ongoing debate over greenhouse gas regulation, and are a major factor in debates over whether to ban particular pesticides. In cases where the science is well-settled, it is not unusual to find that corporations intentionally hide or distort the facts, or sow confusion.

It is very common for regulated industry to argue against environmental regulation on the basis of cost. Difficulties arise in performing cost-benefit analysis of environmental issues. It is difficult to quantify the value of an environmental value such as a healthy ecosystem, clean air, or species diversity. Many environmentalists' response to pitting economy vs. ecology is summed up by former Senator and founder of Earth Day Gaylord Nelson, "The economy is a wholly owned subsidiary of the environment, not the other way around." Furthermore, environmental issues are seen by many as having an ethical or moral dimension, which would transcend financial cost. Even so, there are some efforts underway to systemically recognize environmental costs and assets, and account for them properly in economic terms.

While affected industries spark controversy in fighting regulation, there are also many environmentalists and public interest groups who believe that current regulations are inadequate, and advocate for stronger protection. Environmental law conferences

- such as the annual Public Interest Environmental Law Conference in Eugene, Oregon
- typically have this focus, also connecting environmental law with class, race, and other issues.

An additional debate is to what extent environmental laws are fair to all regulated parties. For instance, researchers Preston Teeter and Jorgen Sandberg highlight how smaller organizations can often incur disproportionately larger costs as a result of environmental regulations, which can ultimately create an additional barrier to entry for new firms, thus stifling competition and innovation.

AROUND THE WORLD

International law

Global and regional environmental issues are increasingly the subject of international law. Debates over environmental concerns implicate core principles of international law and have been the subject of numerous international agreements and declarations.

Customary international law is an important source of international environmental law. These are the norms and rules that countries follow as a matter of custom and they are so prevalent that they bind all states in the world. When a principle becomes customary law is not clear cut and many arguments are put forward by states not wishing to be bound. Examples of customary international law relevant to the environment include the duty to warn other states promptly about icons of an environmental nature and environmental damages to which another state or states may be exposed, and Principle 21 of the Stockholm Declaration ('good neighbourliness' or sic utere).

Numerous legally binding international agreements encompass a wide variety of issue-areas, from terrestrial, marine and atmospheric pollution through to wildlife and biodiversity protection. International environmental agreements are generally multilateral (or sometimes bilateral) treaties (a.k.a. convention, agreement, protocol, etc.). Protocols are subsidiary agreements built from a primary treaty. They exist in many areas of international law but are especially useful in the environmental field, where they may be used to regularly incorporate recent scientific knowledge. They also permit countries to reach agreement on a framework that would be contentious if every detail were to be agreed upon in advance. The most widely known protocol in international environmental law is the Kyoto Protocol, which followed from the United Nations Framework Convention on Climate Change.

While the bodies that proposed, argued, agreed upon and ultimately adopted existing international agreements vary according to each agreement, certain conferences, including 1972's United Nations Conference on the Human Environment, 1983's World Commission on Environment and Development, 1992's United Nations Conference on Environment and Development and 2002's World Summit on Sustainable Development have been particularly important. Multilateral environmental agreements sometimes create an International Organization, Institution or Body responsible for implementing the agreement. Major examples are the Convention on International Trade in Endangered Species of Wild Fauna and Flora (CITES) and the International Union for Conservation of Nature (IUCN).

International environmental law also includes the opinions of international courts and tribunals. While there are few and they have limited authority, the decisions carry much weight with legal commentators and are quite influential on the development of international environmental law. One of the biggest challenges in international decisions is to determine an adequate compensation for environmental damages. The courts include the International Court of Justice (ICJ), the international Tribunal for the Law of the Sea (ITLOS), the European Court of Justice, European Court of Human Rights and other regional treaty tribunals.

Africa

According to the International Network for Environmental Compliance and Enforcement (INECE), the major environmental issues in Africa are "drought and flooding, air pollution, deforestation, loss of biodiversity, freshwater availability, degradation of soil and vegetation, and widespread poverty." The U.S. Environmental Protection Agency (EPA) is focused on the "growing urban and industrial pollution, water quality, electronic waste and indoor air from cookstoves." They hope to provide enough aid on concerns regarding pollution before their impacts contaminate the African environment as well as the global environment. By doing so, they intend to "protect human health, particularly vulnerable populations such as children and the poor." In order to accomplish these goals in Africa, EPA programs are focused on strengthening the ability to enforce environmental laws as well as public compliance to them. Other programs work on developing stronger environmental laws, regulations, and standards.

Asia

The Asian Environmental Compliance and Enforcement Network (AECEN) is an agreement between 16 Asian countries dedicated to improving cooperation with environmental laws in Asia. These countries include Cambodia, China, Indonesia, India, Maldives, Japan, Korea, Malaysia, Nepal, Philippines, Pakistan, Singapore, Sri Lanka, Thailand, Vietnam, and Lao PDR.

European Union

The European Union issues secondary legislation on environmental issues that are valid throughout the EU (so called regulations) and many directives that must be implemented into national legislation from the 28 member states (national states).

Examples are the Regulation (EC) No. 338/97 on the implementation of CITES; or the Natura 2000 network the centerpiece for nature & biodiversity policy, encompassing the bird Directive (79/409/EEC/ changed to 2009/147/EC)and the habitats directive (92/43/EEC). Which are made up of multiple SACs (Special Areas of Conservation, linked to the habitats directive) & SPAs (Special Protected Areas, linked to the bird directive), throughout Europe.

EU legislation is ruled in Article 249 Treaty for the Functioning of the European Union (TFEU). Topics for common EU legislation are:

- Climate change
- Air pollution
- Water protection and management
- Waste management
- Soil protection
- Protection of nature, species and biodiversity
- Noise pollution
- Cooperation for the environment with third countries (other than EU member states)
- Civil protection

Middle East

The U.S. Environmental Protection Agency is working with countries in the Middle East to improve "environmental governance, water pollution and water security , clean fuels and vehicles, public participation, and pollution prevention."

Oceania

The main concerns on environmental issues in the Oceanic Region are "illegal releases of air and water pollutants, illegal logging/timber trade, illegal shipment of hazardous wastes, including e-waste and ships slated for destruction, and insufficient institutional structure/lack of enforcement capacity". The Secretariat of the Pacific Regional Environmental Programme (SPREP) is an international organization between Australia, the Cook Islands, FMS, Fiji, France, Kiribati, Marshall Islands, Nauru, New Zealand, Niue, Palau, PNG, Samoa, Solomon Island, Tonga, Tuvalu, USA, and Vanuatu. The SPREP was established in order to provide assistance in improving and protecting the environment as well as assure sustainable development for future generations.

Australia

The Environment Protection and Biodiversity Conservation Act 1999 is the center piece of environmental legislation in the Australian Government. It sets up the "legal framework to protect and manage nationally and internationally important flora, fauna, ecological communities and heritage places". It also focuses on protecting world heritage properties, national heritage properties, wetlands of international importance, nationally threatened species and ecological communities, migratory species, Commonwealth marine areas, Great Barrier Reef Marine Park, and the environment surrounding nuclear activities. Commonwealth v Tasmania (1983), also known as the "T asmanian Dam Case", is the most influential case for Australian environmental law.

Brazil

The Brazilian government created the Ministry of Environment in 1992 in order to develop better strategies of protecting the environment, use natural resources sustainably, and enforce public environmental policies. The Ministry of Environment has authority over policies involving environment, water resources, preservation, and environmental programs involving the Amazon.

Canada

The Department of the Environment Act establishes the Department of the Environment in the Canadian government as well as the position Minister of the Environment. Their duties include "the preservation and enhancement of the quality of the natural environment, including water, air and soil quality; renewable resources, including migratory birds and other non-domestic flora and fauna; water; meteorology;" The Environmental Protection Act is the main piece of Canadian environmental legislation that was put into place March 31, 2000. The Act focuses on "respecting pollution prevention and the protection of the environment and human health in order to contribute to sustainable development." Other principle federal statutes include the Canadian Environmental Assessment Act, and the Species at Risk Act. When provincial and federal legislation are in conflict federal legislation takes precedence, that being said individual provinces can have their own legislation such as Ontario's Environmental Bill of Rights, and Clean Water Act.

China

According to the U.S. Environmental Protection Agency, "China has been working with great determination in recent years to develop, implement, and enforce a solid environmental law framework. Chinese officials face critical challenges in effectively implementing the laws, clarifying the roles of their national and provincial governments, and strengthening the operation of their legal system." Explosive economic and industrial growth in China has led to significant environmental degradation, and China is currently in the process of developing more stringent legal controls. The harmonization of Chinese society and the natural environment is billed as a rising policy priority.

Ecuador

With the enactment of the 2008 Constitution, Ecuador became the first country in the world to codify the Rights of Nature. The Constitution, specifically Articles 10 and 71-74, recognizes the inalienable rights of ecosystems to exist and flourish, gives people the authority to petition on the behalf of ecosystems, and requires the government to remedy violations of these rights. The rights approach is a break away from traditional environmental regulatory systems, which regard nature as property and legalize and manage degradation of the environment rather than prevent it.

The Rights of Nature articles in Ecuador's constitution are part of a reaction to a combination of political, economic, and social phenomena. Ecuador's abusive past with the oil industry, most famously the class-action litigation against Chevron, and the failure of an extraction-based economy and neoliberal reforms to bring economic prosperity to the region has resulted in the election of a New Leftist regime, led by President Rafael Correa, and sparked a demand for new approaches to development. In conjunction with this need, the principle of "Buen Vivir," or good living— focused on social, environmental and spiritual wealth versus material wealth— gained popularity among citizens and was incorporated into the new constitution.

The influence of indigenous groups, from whom the concept of "Buen Vivir" originates, in the forming of the constitutional ideals also facilitated the incorporation of the Rights of Nature as a basic tenet of their culture and conceptualization of "Buen Vivir."

Egypt

The Environmental Protection Law outlines the responsibilities of the Egyptian government to "preparation of draft legislation and decrees pertinent to environmental management, collection of data both nationally and internationally on the state of the environment, preparation of periodical reports and studies on the state of the environment, formulation of the national plan and its projects, preparation of environmental profiles for new and urban areas, and setting of standards to be used in planning for their development, and preparation of an annual report on the state of the environment to be prepared to the President."

India

In India, Environmental law is governed by the Environment Protection Act, 1986. This act is enforced by the Central Pollution Control Board and the numerous State Pollution Control Boards. Apart from this, there are also individual legislations specifically enacted for the protection of Water, Air, Wildlife, etc. Such legislations include :-

- The Water (Prevention and Control of Pollution) Act, 1974
- The Water (Prevention and Control of Pollution) Cess Act, 1977
- The Forest (Conservation) Act, 1980
- The Air (Prevention and Control of Pollution) Act, 1981
- Air (Prevention and Control of Pollution) (Union Territories) Rules, 1983
- The Biological Diversity Act, 2002 and the Wild Life Protection Act, 1972
- Batteries (Management and Handling) Rules, 2001
- Recycled Plastics, Plastics Manufacture and Usage Rules, 1999
- The National Green Tribunal established under the National Green Tribunal Act of 2010 has jurisdiction over all environmental cases dealing with a substantial environmental question and acts covered under the Water (Prevention and Control of Pollution) Act, 1974.
- Water (Prevention and Control of Pollution) Cess Rules, 1978
- Ganga Action Plan, 1986
- The Forest (Conservation) Act, 1980
- The Public Liability Insurance Act, 1991 and the Biological Diversity Act, 2002. The acts covered under Indian Wild Life Protection Act 1972 do not fall within the jurisdiction of the National Green Tribunal. Appeals can be filed in the Hon'ble Supreme Court of India.
- Basel Convention on Control of TransboundaryMovements on Hazardous Wastes and Their Disposal, 1989 and Its Protocols
- Hazardous Wastes (Management and Handling) Amendment Rules, 2003

Japan

The Basic Environmental Law is the basic structure of Japan's environmental policies replacing the Basic Law for Environmental Pollution Control and the Nature Conservation Law. The updated law aims to address "global environmental problems, urban pollution by everyday life, loss of accessible natural environment in urban areas and degrading environmental protection capacity in forests and farmlands."

The three basic environmental principles that the Basic Environmental Law follows are "the blessings of the environment should be enjoyed by the present generation and succeeded to the future generations, a sustainable society should be created where environmental loads by human activities are minimized, and Japan should contribute actively to global environmental conservation through international cooperation." From these principles, the Japanese government have established policies such as "environmental consideration in policy formulation, establishment of the Basic Environment Plan which describes the directions of long-term environmental policy, environmental impact assessment for development projects, economic measures to encourage activities for reducing environmental load, improvement of social infrastructure such as sewerage system, transport facilities etc., promotion of environmental activities by corporations, citizens and NGOs, environmental education, and provision of information, promotion of science and technology."

New Zealand

were established by the Environment Act 1986. These positions are responsible for advising the Minister on all areas of environmental legislation. A common theme of New Zealand's environmental legislation is sustainably managing natural and physical resources, fisheries, and forests. The Resource Management Act 1991 is the main piece of environmental legislation that outlines the government's strategy to managing the "environment, including air , water soil, biodiversity, the coastal environment, noise, subdivision, and land use planning in general."

Russia

The Ministry of Natural Resources and Environment of the Russian Federation makes regulation regarding "conservation of natural resources, including the subsoil, water bodies, forests located in designated conservation areas, fauna and their habitat, in the field of hunting, hydrometeorology and related areas, environmental monitoring and pollution control, including radiation monitoring and control, and functions of public environmental policy making and implementation and statutory regulation."

Concerns

Environmental health addresses all human-health-related aspects of the natural environment and the built environment. Environmental health concerns include:

- Air quality, including both ambient outdoor air and indoor air quality, which also comprises concerns about environmental tobacco smoke.

- Biosafety
- Climate change and its effects on health.
- Disaster preparedness and response.
- Food safety, including in agriculture, transportation, food processing, wholesale and retail distribution and sale.
- Hazardous materials management, including hazardous waste management, contaminated site remediation, the prevention of leaks from underground storage tanks and the prevention of hazardous materials releases to the environment and responses to emergency situations resulting from such releases.
- Housing, including substandard housing abatement and the inspection of jails and prisons.
- Childhood lead poisoning prevention.
- Land use planning, including smart growth.
- Liquid waste disposal, including city waste water treatment plants and on-site waste water disposal systems, such as septic tank systems and chemical toilets.
- Medical waste management and disposal.
- Noise pollution control.
- Occupational health and industrial hygiene.
- Radiological health, including exposure to ionizing radiation from X-rays or radioactive isotopes.
- Recreational water illness prevention, including from swimming pools, spas and ocean and freshwater bathing places.
- Safe drinking water.
- Solid waste management, including landfills, recycling facilities, composting and solid waste transfer stations.
- Toxic chemical exposure whether in consumer products, housing, workplaces, air, water or soil.
- Vector control, including the control of mosquitoes, rodents, flies, cockroaches and other animals that may transmit pathogens.

According to recent estimates, about 5 to 10% of Disability-adjusted life years (DALYs) lost are due to environmental causes in Europe. By far the most important factor is fine particulate matter pollution in urban air. Similarly, environmental exposures have been estimated to contribute to 4.9 million (8.7%) deaths and 86 million (5.7%) DALYs globally. In the United States, Superfund sites created by various companies have been found to be hazardous to human and environmental health in nearby communities. It was this perceived threat, raising the specter of miscarriages, mutations, birth defects, and cancers that most frightened the public.

Information

The Toxicology and Environmental Health Information Program (TEHIP) is a comprehensive toxicology and environmental health web site, that includes open access to resources produced by US government agencies and organizations, and is maintained under the umbrella of the Specialized Information Service at the United States National Library of Medicine. TEHIP includes links to technical databases, bibliographies, tutorials, and consumer-oriented resources. TEHIP is responsible for the Toxicology Data Network (TOXNET), an integrated system of toxicology and environmental health databases including the Hazardous Substances Data Bank, that are open access, i.e. available free of charge.

Mapping

There are many environmental health mapping tools. TOXMAP is a geographic information system (GIS) from the Division of Specialized Information Services of the United States National Library of Medicine (NLM) that uses maps of the United States to help users visually explore data from the United States Environmental Protection Agency's (EPA) Toxics Release Inventory and Superfund Basic Research Programs. TOXMAP is a resource funded by the US federal government. TOXMAP's chemical and environmental health information is taken from the NLM's Toxicology Data Network (TOXNET) and PubMed, and from other authoritative sources.

4

Evolution and Biodiversity

Biodiversity, a portmanteau of "bio" (life) and "diversity", generally refers to the variety and variability of life on Earth. According to the United Nations Environment Programme (UNEP), biodiversity typically measures variation at the genetic, the species, and the ecosystem level. Terrestrial biodiversity tends to be greater near the equator, which seems to be the result of the warm climate and high primary productivity. Biodiversity is not distributed evenly on Earth, and is richest in the tropics. These tropical forest ecosystems cover less than 10 percent of earth's surface, and contain about 90 percent of the world's species. Marine biodiversity tends to be highest along coasts in the Western Pacific, where sea surface temperature is highest and in the mid-latitudinal band in all oceans. There are latitudinal gradients in species diversity. Biodiversity generally tends to cluster in hotspots, and has been increasing through time, but will be likely to slow in the future.

Rapid environmental changes typically cause mass extinctions. More than 99.9 percent of all species that ever lived on Earth, amounting to over five billion species, are estimated to be extinct. Estimates on the number of Earth's current species range from 10 million to 14 million, of which about 1.2 million have been documented and over 86 percent have not yet been described. More recently, in May 2016, scientists reported that 1 trillion species are estimated to be on Earth currently with only one-thousandth of one percent described. The total amount of related DNA base pairs on Earth is estimated at 5.0 x 10 and weighs 50 billion tonnes. In comparison, the total mass of the biosphere has been estimated to be as much as 4 TtC (trillion tons of carbon). In July 2016, scientists reported identifying a set of 355 genes from the Last Universal Common Ancestor (LUCA) of all organisms living on Earth.

The age of the Earth is about 4.54 billion years. The earliest undisputed evidence of life on Earth dates at least from 3.5 billion years ago, during the Eoarchean Era after a geological crust started to solidify following the earlier molten Hadean Eon. There are microbial mat fossils found in 3.48 billion-year-old sandstone discovered in Western Australia. Other early physical evidence of a biogenic substance is graphite in 3.7 billion-year-old meta-sedimentary rocksdiscovered in Western Greenland. More recently, in 2015, "remains of biotic life" were found in 4.1 billion-year -old rocks in Western Australia. According to one of the researchers, "If life arose relatively quickly on Earth .. then it could be common in the universe."

Since life began on Earth, five major mass extinctions and several minor events have led to large and sudden drops in biodiversity. The Phanerozoic eon (the last 540 million years) marked a rapid growth in biodiversity via the Cambrian explosion—a period during which the majority of multicellular phyla first appeared. The next 400 million years included repeated, massive biodiversity losses classified as mass extinction events. In the Carboniferous, rainforest collapse led to a great loss of plant and animal life. The Permian–Triassic extinction event, 251 million years ago, was the worst; vertebrate recovery took 30 million years. The most recent, the Cretaceous–Paleogene extinction event, occurred 65 million years ago and has often attracted more attention than others because it resulted in the extinction of the dinosaurs.

The period since the emergence of humans has displayed an ongoing biodiversity reduction and an accompanying loss of genetic diversity. Named the Holocene extinction, the reduction is caused primarily by human impacts, particularly habitat destruction. Conversely, biodiversity positively impacts human health in a number of ways, although a few negative effects are studied.

The United Nations designated 2011–2020 as the United Nations Decade on Biodiversity.

ETYMOLOGY

The term biological diversity was used first by wildlife scientist and conservationist Raymond F. Dasmann in the year 1968 lay book A Different Kind of Country advocating conservation. The term was widely adopted only after more than a decade, when in the 1980s it came into common usage in science and environmental policy. Thomas Lovejoy, in the foreword to the book Conservation Biology, introduced the term to the scientific community. Until then the term "natural diversity" was common, introduced by The Science Division of The Nature Conservancy in an important 1975 study, "The Preservation of Natural Diversity." By the early 1980s TNC's Science program and its head, Robert E. Jenkins, Lovejoy and other leading conservation scientists at the time in America advocated the use of the term "biological diversity".

The term's contracted form biodiversity may have been coined by W.G. Rosen in 1985 while planning the 1986 National Forum on Biological Diversity organized by the National Research Council (NRC). It first appeared in a publication in 1988 when sociobiologist E. O. Wilson used it as the title of the proceedings of that forum.

Since this period the term has achieved widespread use among biologists, environmentalists, political leaders and concerned citizens.

A similar term in the United States is "natural heritage." It pre-dates the others and is more accepted by the wider audience interested in conservation. Broader than biodiversity, it includes geology and landforms.

Definitions

A sampling of fungi collected during summer 2008 in Northern Saskatchewan mixed woods, near LaRonge is an example regarding the species diversity of fungi. In this photo, there are also leaf lichens and mosses.

"Biodiversity" is most commonly used to replace the more clearly defined and long established terms, species diversity and species richness. Biologists most often define biodiversity as the "totality of genes, species and ecosystems of a region". An advantage of this definition is that it seems to describe most circumstances and presents a unified view of the traditional types of biological variety previously identified:

- taxonomic diversity (usually measured at the species diversity level)
- ecological diversity often viewed from the perspective of ecosystem diversity
- morphological diversity which stems from genetic diversityand molecular diversity

• functional diversity which is a measure of the number of functionally disparate species within a population (e.g. different feeding mechanism, different motility, predator vs prey, etc.)

This multilevel construct is consistent with Datman and Lovejoy. An explicit definition consistent with this interpretation was first given in a paper by Bruce A. Wilcox commissioned by the International Union for the Conservation of Nature and Natural Resources (IUCN) for the 1982 World National Parks Conference. Wilcox's definition was "Biological diversity is the variety of life forms...at all levels of biological systems (i.e., molecular, organismic, population, species and ecosystem)...". The 1992 United Nations Earth Summitdefined "biological diversity" as "the variability among living organisms from all sources, including, 'inter alia', terrestrial, marine and other aquatic ecosystems and the ecological complexes of which they are part: this includes diversity within species, between species and of ecosystems". This definition is used in the United Nations Convention on Biological Diversity.

One textbook's definition is "variation of life at all levels of biological or ganization". Biodiversity can be defined genetically as the diversity of alleles, genes and organisms. They study processes such as mutation and gene transfer that drive evolution. Measuring diversity at one level in a group of organisms may not precisely correspond to diversity at other levels. However, tetrapod (terrestrial vertebrates) taxonomic and ecological diversity shows a very close correlation.

Distribution

A conifer forest in the Swiss Alps(National Park)

Biodiversity is not evenly distributed, rather it varies greatly across the globe as well as within regions. Among other factors, the diversity of all living things (biota) depends on temperature, precipitation, altitude, soils, geography and the presence of other species. The study of the spatial distribution of organisms, species and ecosystems, is the science of biogeography.

Diversity consistently measures higher in the tropics and in other localized regions such as the Cape Floristic Region and lower in polar regions generally. Rain forests that have had wet climates for a long time, such as Yasuní National Park in Ecuador, have particularly high biodiversity.

Terrestrial biodiversity is thought to be up to 25 times greater than ocean biodiversity. A recently discovered method put the total number of species on Earth at 8.7 million, of which 2.1 million were estimated to live in the ocean. However, this estimate seems to under-represent the diversity of microorganisms.

Latitudinal gradients

Generally, there is an increase in biodiversity from the poles to the tropics. Thus localities at lower latitudes have more species than localities at higher latitudes. This is often referred to as the latitudinal gradient in species diversity. Several ecological mechanisms may contribute to the gradient, but the ultimate factor behind many of them is the greater mean temperature at the equator compared to that of the poles.

Even though terrestrial biodiversity declines from the equator to the poles, some studies claim that this characteristic is unverified in aquatic ecosystems, especially in marine ecosystems. The latitudinal distribution of parasites does not appear to follow this rule.

In 2016, an alternative hypothesis ("the fractal biodiversity") was proposed to explain the biodiversity latitudinal gradient . In this study, the species pool size and the fractal nature of ecosystems were combined to clarify some general patterns of this gradient. This hypothesis considers temperature, moisture, and net primary production (NPP) as the main variables of an ecosystem niche and as the axis of the ecological hypervolume. In this way, it is possible to build fractal hypervolumes, whose fractal dimension rises up to three moving towards the equator.

Hotspots

A biodiversity hotspot is a region with a high level of endemic species that has experienced great habitat loss. The term hotspot was introduced in 1988 by Norman Myers. While hotspots are spread all over the world, the majority are forest areas and most are located in the tropics.

Brazil's Atlantic Forest is considered one such hotspot, containing roughly 20,000 plant species, 1,350 vertebrates and millions of insects, about half of which occur nowhere else. The island of Madagascar and India are also particularly notable. Colombia is characterized by high biodiversity, with the highest rate of species by area unit worldwide and it has the largest number of endemics (species that are not found naturally anywhere else) of any country. About 10% of the species of the Earth can be found in Colombia, including over 1,900 species of bird, more than in Europe and North America combined, Colombia has 10% of the world's mammals species, 14% of the amphibian species and 18% of the bird species of the world. Madagascar dry deciduous forests and lowland rainforests possess a high ratio of endemism. Since the island separated from mainland Africa 66 million years ago, many species and ecosystems have evolved independently. Indonesia's 17,000 islands cover 735,355 square miles (1,904,560 km) and contain 10% of the world's flowering plants, 12% of mammals and 17% of reptiles, amphibians and birds—along with nearly 240 million people.

Many regions of high biodiversity and/or endemism arise from specialized habitatswhich require unusual adaptations, for example, alpine environments in high mountains, or Northern European peat bogs.

Accurately measuring differences in biodiversity can be difficult. Selection bias amongst researchers may contribute to biased empirical research for modern estimates of biodiversity. In 1768, Rev. Gilbert White succinctly observed of his Selborne, Hampshire "all nature is so full, that that district produces the most variety which is the most examined."

EVOLUTION AND HiSTORY

Evolution is change in the heritable characteristics of biological populations over successive generations.Evolutionary processes give rise to biodiversity at every level of biological organisation, including the levels of species, individual organisms, and molecules.

Repeated formation of new species (speciation), change within species (anagenesis), and loss of species (extinction) throughout the evolutionary history of life on Earth are demonstrated by shared sets of morphological and biochemicaltraits, including shared DNA sequences. These shared traits are more similar among species that share a more recent common ancestor, and can be used to reconstruct a biological "tree of life" based on evolutionary relationships (phylogenetics), using both existing species and fossils. The fossil record includes a progression from early biogenicgraphite, to microbial mat fossils, to fossilised multicellular organisms. Existing patterns of biodiversity have been shaped both by speciation and by extinction.

In the mid-19th century, Charles Darwin formulated the scientific theory of evolution by natural selection, published in his book On the Origin of Species (1859). Evolution by natural selection is a process demonstrated by the observation that more offspring are produced than can possibly survive, along with three facts about populations: 1) traits vary among individuals with respect to morphology, physiology, and behaviour (phenotypic variation), 2) different traits confer different rates of survival and reproduction (differential fitness), and 3) traits can be passed from generation to generation (heritability of fitness). Thus, in successive generations members of a population are replaced by progeny of parents better adapted to survive and reproduce in the biophysical environment in which natural selection takes place.

This teleonomy is the quality whereby the process of natural selection creates and preserves traits that are seemingly fitted for the functional roles they perform. The processes by which the changes occur, from one generation to another, are called evolutionary processes or mechanisms. The four most widely recognised evolutionary processes are natural selection (including sexual selection), genetic drift, mutation and gene migration due to genetic admixture. Natural selection and genetic drift sort variation; mutation and gene migration create variation.

Consequences of selection can include meiotic drive (unequal transmission of certain alleles), nonrandom mating and genetic hitchhiking. In the early 20th century the modern evolutionary synthesis integrated classical genetics with Darwin's theory of evolution by natural selection through the discipline of population genetics. The importance of natural selection as a cause of evolution was accepted into other branches of biology. Moreover, previously held notions about evolution, such as orthogenesis, evolutionism, and other beliefs about innate "progress" within the largest-scale trends in evolution, became obsolete. Scientists continue to study various aspects of evolutionary biology by forming and testing hypotheses, constructing mathematical models of theoretical biology and biological theories, using observational data, and performing experiments in both the field and the laboratory.

All life on Earth shares a common ancestor known as the last universal common ancestor (LUCA), which lived approximately 3.5–3.8 billion years ago. A December 2017 report stated that 3.45 billion year old Australian rocks once contained microorganisms, the earliest direct evidence of life on Earth. Nonetheless, this should not be assumed to be the first living organism on Earth; a study in 2015 found "remains of biotic life" from 4.1 billion years ago in ancient rocks in Western Australia. In July 2016, scientists reported identifying a set of 355 genes from the LUCA of all organisms living on Earth. More than 99 percent of all species that ever lived on Earth are estimated to be extinct. Estimates of Earth's current species range from 10 to 14 million, of which about 1.9 million are estimated to have been named and 1.6 million documented in a central database to date. More recently, in May 2016, scientists reported that 1 trillion species are estimated to be on Earth currently with only one-thousandth of one percent described.

In terms of practical application, an understanding of evolution has been instrumental to developments in numerous scientific and industrial fields, including agriculture, human and veterinary medicine, and the life sciences in general. Discoveries in evolutionary biology have made a significant impact not just in the traditional branches of biology but also in other academic disciplines, including biological anthropology, and evolutionary psychology. Evolutionary computation, a sub-field of artificial intelligence, involves the application of Darwinian principles to problems in computer science.

The diversity of the living world is staggering. More than 2 million existing species of organisms have been named and described; many more remain to be discovered— from 10 million to 30 million, according to some estimates. What is impressive is not just the numbers but also the incredible heterogeneity in size, shape, and way of life— from lowly bacteria, measuring less than a thousandth of a millimetre in diameter, to stately sequoias, rising 100 metres (300 feet) above the ground and weighing several thousand tons; from bacteria living in hot springs at temperatures near the boiling point of water to fungi and algae thriving on the ice masses of Antarctica and in saline pools at "23 °C ("9 °F); and from giant tube worms discovered living near hydrothermal vents on the dark ocean floor to spiders and larkspur plants existing on the slopes of Mount Everest more than 6,000 metres (19,700 feet) above sea level.

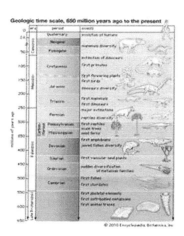

The geologic time scale from 650 million years ago to the present, showing major evolutionary events.Encyclopædia Britannica, Inc.

The virtually infinite variations on life are the fruit of the evolutionary process. All living creatures are related by descent from common ancestors. Humans and other mammals descend from shrewlike creatures that lived more than 150 million years ago; mammals, birds, reptiles, amphibians, and fishes share as ancestors aquatic worms that lived 600 million years ago; and all plants and animals derive from bacteria-like microorganisms that originated more than 3 billion years ago. Biological evolution is a process of descent with modification. Lineages of organisms change through generations; diversity arises because the lineages that descend from common ancestors diverge through time.

The 19th-century English naturalist Charles Darwin argued that organisms come about by evolution, and he provided a scientific explanation, essentially correct but incomplete, of how evolution occurs and why it is that organisms have features—such as wings, eyes, and kidneys—clearly structured to serve specific functions. Natural selectionwas the fundamental concept in his explanation. Natural selection occurs because individuals having more-useful traits, such as more-acute vision or swifter legs, survive better and produce more progeny than individuals with less-favourable traits. Genetics, a science born in the 20th century, reveals in detail how natural selection works and led to the development of the modern theory of evolution. Beginning in the 1960s, a related scientific discipline, molecular biology, enormously advanced knowledge of biological evolution and made it possible to investigate detailed problems that had seemed completely out of reach only a short time previously—for example, how similar the genes of humans and chimpanzees might be (they differ in about 1–2 percent of the units that make up the genes).

This article discusses evolution as it applies generally to living things. For a discussion of human evolution, see the article human evolution. For a more complete treatment of a discipline that has proved essential to the study of evolution, see the articles genetics, human and heredity. Specific aspects of evolution are discussed in the articles coloration and mimicry. Applications of evolutionary theory to plant and animal breeding are discussed in the articles plant breeding and animal breeding. An overview of the evolution of life as a major characteristic of Earth's history is given in community ecology: Evolution of the biosphere. A detailed discussion of the life and thought of Charles Darwin is found in the article Darwin, Charles.

GENERAL OVERVIEW

The evidence for evolution

Darwin and other 19th-century biologists found compelling evidence for biological evolution in the comparative study of living organisms, in their geographic distribution, and in the fossil remains of extinct organisms. Since Darwin's time, the evidence from these sources has become considerably stronger and more comprehensive, while biological disciplines that emerged more recently—genetics, biochemistry, physiology, ecology, animal behaviour (ethology), and especially molecular biology—have supplied powerful additional evidence and detailed confirmation. The amount of information about evolutionary history stored in the DNA and proteins of living things is virtually unlimited; scientists can reconstruct any detail of the evolutionary history of life by investing sufficient time and laboratory resources.

Evolutionists no longer are concerned with obtaining evidence to support the fact of evolution but rather are concerned with what sorts of knowledge can be obtained from different sources of evidence. The following sections identify the most productive of these sources and illustrate the types of information they have provided.

The fossil record

Paleontologists have recovered and studied the fossil remains of many thousands of organisms that lived in the past. This fossil record shows that many kinds of extinct organisms were very different in form from any now living. It also shows successions of organisms through time (see faunal succession, law of; geochronology: Determining the relationships of fossils with rock strata), manifesting their transition from one form to another.

When an organism dies, it is usually destroyed by other forms of life and by weathering processes. On rare occasions some body parts—particularly hard ones such as shells, teeth, or bones—are preserved by being buried in mud or protected in some other way from predators and weather. Eventually, they may become petrified and preserved indefinitely with the rocks in which they are embedded. Methods such as radiometric dating—measuring the amounts of natural radioactive atoms that remain in certain minerals to determine the elapsed time since they were constituted—make it possible to estimate the time period when the rocks, and the fossils associated with them, were formed.

Radiometric dating indicates that Earth was formed about 4.5 billion years ago. The earliest fossils resemble microorganisms such as bacteria and cyanobacteria (blue-green algae); the oldest of these fossils appear in rocks 3.5 billion years old (see Precambrian time). The oldest known animal fossils, about 700 million years old, come from the so-called Ediacara fauna, small wormlike creatures with soft bodies. Numerous fossils belonging to many living phyla and exhibiting mineralized skeletons appear in rocks about 540 million years old. These organisms are different from organisms living now and from those living at intervening times. Some are so radically different that paleontologists have created new phyla in order to classify them. (SeeCambrian Period.) The first vertebrates, animals with backbones, appeared about 400 million years ago; the first mammals, less than 200 million years ago. The history of life recorded by fossils presents compelling evidence of evolution.

The fossil record is incomplete. Of the small proportion of organisms preserved as fossils, only a tiny fraction have been recovered and studied by paleontologists. In some cases the succession of forms over time has been reconstructed in detail. One example is the evolution of the horse. The horse can be traced to an animal the size of a dog having several toes on each foot and teeth appropriate for browsing; this animal, called the dawn horse (genus Hyracotherium), lived more than 50 million years ago. The most recent form, the modern horse (Equus), is much larger in size, is one-toed, and has teeth appropriate for grazing. The transitional forms are well preserved as fossils, as are many other kinds of extinct horses that evolved in different directions and left no living descendants.

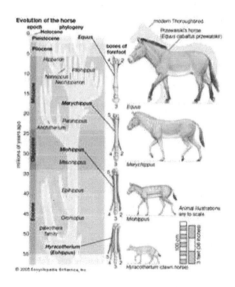

evolution of the horseEvolution of the horse over the past 55 million years. The present-day Przewalski's horse is believed to be the only remaining example of a wild horse—i.e., the last remaining modern horse to have evolved by natural selection.

Numbered bones in the forefoot illustrations trace the gradual transition from a four-toed to a one-toed animal.Encyclopædia Britannica, Inc.

Using recovered fossils, paleontologists have reconstructed examples of radical evolutionary transitions in form and function. For example, the lower jaw of reptiles contains several bones, but that of mammals only one. The other bones in the reptile jaw unmistakably evolved into bones now found in the mammalian ear. At first, such a transition would seem unlikely— it is hard to imagine what function such bones could have had during their intermediate stages. Yet paleontologists discovered two transitional forms of mammal-like reptiles, called therapsids, that had a double jaw joint (i.e., two hinge points side by side)—one joint consisting of the bones that persist in the mammalian jaw and the other composed of the quadrate and articular bones, which eventually became the hammer and anvil of the mammalian ear. (See also mammal: Skeleton.)

For skeptical contemporaries of Darwin, the "missing link"—the absence of any known transitional form between apes and humans—was a battle cry , as it remained for uninformed people afterward. Not one but many creatures intermediate between living apes and humans have since been found as fossils. The oldest known fossil hominins—i.e., primates belonging to the human lineage after it separated from lineages going to the apes— are 6 million to 7 million years old, come from Africa, and are known as Sahelanthropus and Orrorin (or Praeanthropus), which were predominantly bipedal when on the ground but which had very small brains. Ardipithecus lived about 4.4 million years ago, also in Africa. Numerous fossil remains from diverseAfrican origins are known of Australopithecus, a hominin that appeared between 3 million and 4 million years ago. Australopithecushad an upright human stance but a cranial capacity of less than 500 cc (equivalent to a brain weight of about 500 grams), comparable to that of a gorilla or a chimpanzee and about one-third that of humans. Its head displayed a mixture of ape and human characteristics—a low forehead and a long, apelike face but with teeth proportioned like those of humans. Other early hominins partly contemporaneous with Australopithecus include Kenyanthropus and Paranthropus; both had comparatively small brains, although some species of Paranthropushad larger bodies. Paranthropus represents a side branch in the hominin lineage that became extinct. Along with increased cranial capacity, other human characteristics have been found in Homo habilis, which lived about 1.5 million to 2 million years ago in Africa and had a cranial capacity of more than 600 cc (brain weight of 600 grams), and in H. erectus, which lived between 0.5 million and more than 1.5 million years ago, apparently ranged widely over Africa, Asia, and Europe, and had a cranial capacity of 800 to 1,100 cc (brain weight of 800 to 1,100 grams). The brain sizes of H. ergaster, H. antecessor, and H.

heidelbergensis were roughly that of the brain of H. erectus, some of which species were partly contemporaneous, though they lived in different regions of the Eastern Hemisphere. (See also human evolution.)

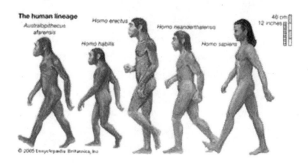

Five hominins—members of the human lineage after it separated at least seven million to six million years ago from lineages going to the apes—are depicted in an artist' s interpretations. All but Homo sapiens, the species that comprises modern humans, are extinct and have been reconstructed from fossil evidence.Encyclopædia Britannica, Inc.

Structural similarities

The skeletons of turtles, horses, humans, birds, and bats are strikingly similar, in spite of the different ways of life of these animals and the diversity of their environments. The correspondence, bone by bone, can easily be seen not only in the limbs but also in every other part of the body. From a purely practical point of view, it is incomprehensible that a turtle should swim, a horse run, a person write, and a bird or a bat flywith forelimb structures built of the same bones. An engineer could design better limbs in each case. But if it is accepted that all of these skeletons inherited their structures from a common ancestor and became modified only as they adapted to different ways of life, the similarity of their structures makes sense.

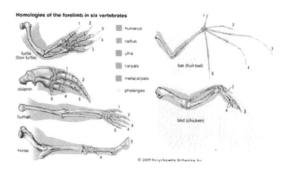

Homologies of the forelimb among vertebrates, giving evidence for evolution. The bones correspond, although they are adapted to the specific mode of life of the animal. (Some anatomists interpret the digits in the bird's wing as being 1, 2, and 3, rather than 2, 3, and 4.)Encyclopædia Britannica, Inc.

Comparative anatomy investigates the homologies, or inherited similarities, among organisms in bone structure and in other parts of the body. The correspondence of structures is typically very close among some organisms—the different varieties of songbirds, for instance—but becomes less so as the organisms being compared are less closely related in their evolutionary history. The similarities are less between mammals and birds than they are among mammals, and they are still less between mammals and fishes. Similarities in structure, therefore, not only manifest evolution but also help to reconstruct the phylogeny, or evolutionary history, of organisms.

Comparative anatomy also reveals why most organismic structures are not perfect. Like the forelimbs of turtles, horses, humans, birds, and bats, an organism's body parts are less than perfectly adapted because they are modified from an inherited structure rather than designed from completely "raw" materials for a specific purpose. The imperfection of structures is evidence for evolution and against antievolutionist arguments that invoke intelligent design (see belowIntelligent design and its critics).

Embryonic development and vestiges

Darwin and his followers found support for evolution in the study of embryology, the science that investigates the development of organisms from fertilized egg to time of birth or hatching. Vertebrates, from fishes through lizards to humans, develop in ways that are remarkably similar during early stages, but they become more and more differentiated as the embryos approach maturity. The similarities persist longer between organisms that are more closely related (e.g., humans and monkeys) than between those less closely related (humans and sharks). Common developmental patterns reflect evolutionary kinship. Lizards and humans share a developmental pattern inherited from their remote common ancestor; the inherited pattern of each was modified only as the separate descendant lineages evolved in different directions. The common embryonic stages of the two creatures reflect the constraints imposed by this common inheritance, which prevents changes that have not been necessitated by their diverging environments and ways of life.

The embryos of humans and other nonaquatic vertebrates exhibit gill slits even though they never breathe through gills. These slits are found in the embryos of all vertebrates because they share as common ancestors the fish in which these structures first evolved. Human embryos also exhibit by the fourth week of development a well-defined tail, which reaches maximum length at six weeks. Similar embryonic tails are found in other mammals, such as dogs, horses, and monkeys; in humans, however, the tail eventually shortens, persisting only as a rudiment in the adult coccyx.

A close evolutionary relationship between organisms that appear drastically different as adults can sometimes be recognized by their embryonic homologies. Barnacles, for example, are sedentary crustaceans with little apparent likeness to such free-swimming crustaceans as lobsters, shrimps, or copepods. Yet barnacles pass through a free-swimming larval stage, the nauplius, which is unmistakably similar to that of other crustacean larvae.

Embryonic rudiments that never fully develop, such as the gill slits in humans, are common in all sorts of animals. Some, however, like the tail rudiment in humans, persist as adult vestiges, reflecting evolutionary ancestry. The most familiar rudimentary organ in humans is the vermiform appendix. This wormlike structure attaches to a short section of intestine called the cecum, which is located at the point where the large and small intestines join. The human vermiform appendix is a functionless vestige of a fully developed organ present in other mammals, such as the rabbit and other herbivores, where a large cecum and appendix store vegetable cellulose to enable its digestionwith the help of bacteria. Vestiges are instances of imperfections—like the imperfections seen in anatomical structures—that argue against creation by design but are fully understandable as a result of evolution.

Biogeography

Darwin also saw a confirmation of evolution in the geographic distribution of plants and animals, and later knowledge has reinforced his observations. For example, there are about 1,500 known species of Drosophila vinegar flies in the world; nearly one-third of them live in Hawaii and nowhere else, although the total area of the archipelago is less than one-twentieth the area of California or Germany. Also in Hawaii are more than 1,000 species of snails and other land mollusks that exist nowhere else. This unusual diversity is easily explained by evolution. The islands of Hawaii are extremely isolated and have had few colonizers—i.e, animals and plants that arrived there from elsewhere and established populations. Those species that did colonize the islands found many unoccupied ecological niches, local environments suited to sustaining them and lacking predators that would prevent them from multiplying. In response, these species rapidly diversified; this process of diversifying in order to fill ecological niches is called adaptive radiation.

Each of the world's continents has its own distinctive collection of animals and plants. In Africa are rhinoceroses, hippopotamuses, lions, hyenas, giraffes, zebras, lemurs, monkeys with narrow noses and nonprehensile tails, chimpanzees, and gorillas. South America, which extends over much the same latitudes as Africa, has none of these animals; it instead has pumas, jaguars, tapir, llamas, raccoons, opossums, armadillos, and monkeys with broad noses and large prehensile tails.

These vagaries of biogeography are not due solely to the suitability of the different environments. There is no reason to believe that South American animals are not well suited to living in Africa or those of Africa to living in South America. The islands of Hawaii are no better suited than other Pacific islands for vinegar flies, nor are they less hospitable than other parts of the world for many absent organisms. In fact, although no large mammals are native to the Hawaiian islands, pigs and goats have multiplied there as wild animals since being introduced by humans. This absence of many species from a hospitable environment in which an extraordinary variety of other species flourish can be explained by the theory of evolution, which holds that species can exist and evolve only in geographic areas that were colonized by their ancestors.

Molecular biology

The field of molecular biology provides the most detailed and convincing evidence available for biological evolution. In its unveiling of the nature of DNA and the workings of organisms at the level of enzymes and other protein molecules, it has shown that these molecules hold information about an organism's ancestry. This has made it possible to reconstruct evolutionary events that were previously unknown and to confirm and adjust the view of events already known. The precision with which these events can be reconstructed is one reason the evidence from molecular biology is so compelling. Another reason is that molecular evolution has shown all living organisms, from bacteria to humans, to be related by descent from common ancestors.

A remarkable uniformity exists in the molecular components of organisms—in the nature of the components as well as in the ways in which they are assembled and used. In all bacteria, plants, animals, and humans, the DNA comprises a different sequence of the same four component nucleotides, and all the various proteins are synthesized from different combinations and sequences of the same 20 amino acids, although several hundred other amino acids do exist. The genetic code by which the information contained in the DNA of the cellnucleus is passed on to proteins is virtually everywhere the same. Similar metabolic pathways—sequences of biochemical reactions (seemetabolism)—are used by the most diverse organisms to produce energy and to make up the cell components.

This unity reveals the genetic continuity and common ancestry of all organisms. There is no other rational way to account for their molecular uniformity when numerous alternative structures are equally likely. The genetic code serves as an example. Each particular sequence of three nucleotides in the nuclear DNA acts as a pattern for the production of exactly the same amino acid in all organisms. This is no more necessary than it is for a language to use a particular combination of letters to represent a particular object. If it is found that certain sequences of letters—planet, tree, woman— are used with identical meanings in a number of different books, one can be sure that the languages used in those books are of common origin.

Genes and proteins are long molecules that contain information in the sequence of their components in much the same way as sentences of the English language contain information in the sequence of their letters and words. The sequences that make up the genes are passed on from parents to offspring and are identical except for occasional changes introduced by mutations. As an illustration, one may assume that two books are being compared. Both books are 200 pages long and contain the same number of chapters. Closer examination reveals that the two books are identical page for page and word for word, except that an occasional word—say , one in 100—is dif ferent. The two books cannot have been written independently; either one has been copied from the other, or both have been copied, directly or indirectly, from the same original book. Similarly, if each component nucleotide of DNA is represented by one letter, the complete sequence of nucleotides in the DNA of a higher organism would require several hundred books of hundreds of pages, with several thousand letters on each page. When the "pages" (or sequences of nucleotides) in these "books" (organisms) are examined one by one, the correspondence in the "letters" (nucleotides) gives unmistakable evidence of common origin.

The two arguments presented above are based on different grounds, although both attest to evolution. Using the alphabet analogy, the first argument says that languages that use the same dictionary—the same genetic code and the same 20 amino acids— cannot be of independent origin. The second argument, concerning similarity in the sequence of nucleotides in the DNA (and thus the sequence of amino acids in the proteins), says that books with very similar texts cannot be of independent origin.

The evidence of evolution revealed by molecular biology goes even farther. The degree of similarity in the sequence of nucleotides or of amino acids can be precisely quantified. For example, in humans and chimpanzees, the protein molecule called cytochrome c, which serves a vital function in respiration within cells, consists of the same 104 amino acids in exactly the same order. It differs, however, from the cytochrome c of rhesus monkeys by 1 amino acid, from that of horses by 11 additional amino acids, and from that of tuna by 21 additional amino acids. The degree of similarity reflects the recency of common ancestry. Thus, the inferences from comparative anatomy and other disciplines concerning evolutionary history can be tested in molecular studies of DNA and proteins by examining their sequences of nucleotides and amino acids. (See below DNA and protein as informational macromolecules.)

The authority of this kind of test is overwhelming; each of the thousands of genes and thousands of proteins contained in an organism provides an independent test of that organism's evolutionary history. Not all possible tests have been performed, but many hundreds have been done, and not one has given evidence contrary to evolution. There is probably no other notion in any field of science that has been as extensively tested and as thoroughly corroborated as the evolutionary origin of living organisms.

HISTORY OF EVOLUTIONARY THEORY

Early ideas

All human cultures have developed their own explanations for the origin of the world and of human beings and other creatures. Traditional Judaism and Christianity explain the origin of living beings and their adaptations to their environments—wings, gills, hands, flowers—as the handiwork of an omniscient God. The philosophers of ancient Greece had their own creation myths. Anaximander proposed that animals could be transformed from one kind into another, and Empedocles speculated that they were made up of various combinations of preexisting parts. Closer to modern evolutionary ideas were the proposals of early Church Fathers such as Gregory of Nazianzus and Augustine, both of whom maintained that not all species of plants and animals were created by God; rather, some had developed in historical times from God's creations. Their motivation was not biological but religious—it would have been impossible to hold representatives of all species in a single vessel such as Noah's Ark; hence, some species must have come into existence only after the Flood.

The notion that organisms may change by natural processes was not investigated as a biological subject by Christian theologians of the Middle Ages, but it was, usually incidentally, considered as a possibility by many, including Albertus Magnus and his student Thomas Aquinas. Aquinas concluded, after detailed discussion, that the development of living creatures such as maggots and flies from nonliving matter such as decaying meat was not incompatible with Christian faith or philosophy. But he left it to others to determine whether this actually happened.

The idea of progress, particularly the belief in unbounded human progress, was central to the Enlightenment of the 18th century, particularly in France among such philosophers as the marquis de Condorcet and Denis Diderot and such scientists as Georges-Louis Leclerc, comte de Buffon. But belief in progress did not necessarily lead to the development of a theory of evolution. Pierre-Louis Moreau de Maupertuis proposed the spontaneous generation and extinction of organisms as part of his theory of origins, but he advanced no theory of evolution—i.e., the transformation of one species into another through knowable, natural causes. Buffon, one of the greatest naturalists of the time, explicitly considered—and rejected—the possible descent of several species from a common ancestor. He postulated that organisms arise from organic molecules by spontaneous generation, so that there could be as many kinds of animals and plants as there are viable combinations of organic molecules.

The English physician Erasmus Darwin, grandfather of Charles Darwin, offered in his Zoonomia; or, The Laws of Organic Life (1794–96) some evolutionary speculations, but they were not further developed and had no real influence on subsequent theories. The Swedish botanist Carolus Linnaeus devised the hierarchical system of plant and animal classification that is still in use in a modernized form. Although he insisted on the fixity of species, his classification system eventually contributed much to the acceptance of the concept of common descent.

The great French naturalist Jean-Baptiste de Monet, chevalier de Lamarck, held the enlightened view of his age that living organisms represent a progression, with

humans as the highest form. From this idea he proposed, in the early years of the 19th century, the first broad theory of evolution. Organisms evolve through eons of time from lower to higher forms, a process still going on, always culminating in human beings. As organisms become adapted to their environments through their habits, modifications occur. Use of an organ or structure reinforces it; disuse leads to obliteration. The characteristics acquired by use and disuse, according to this theory, would be inherited. This assumption, later called the inheritance of acquired characteristics (or Lamarckism), was thoroughly disproved in the 20th century. Although his theory did not stand up in the light of later knowledge, Lamarck made important contributions to the gradual acceptance of biological evolution and stimulated countless later studies.

Charles Darwin

The founder of the modern theory of evolution was Charles Darwin. The son and grandson of physicians, he enrolled as a medical student at the University of Edinburgh. After two years, however, he left to study at the University of Cambridge and prepare to become a clergyman. He was not an exceptional student, but he was deeply interested in natural history. On December 27, 1831, a few months after his graduation from Cambridge, he sailed as a naturalist aboard the HMS Beagle on a round-the-world trip that lasted until October 1836. Darwin was often able to disembark for extended trips ashore to collect natural specimens.

The discovery of fossil bones from large extinct mammals in Argentina and the observation of numerous species of finches in the Galapagos Islands were among the events credited with stimulating Darwin's interest in how species originate. In 1859 he published On the Origin of Species by Means of Natural Selection, a treatise establishing the theory of evolution and, most important, the role of natural selection in determining its course. He published many other books as well, notably The Descent of Man and Selection in Relation to Sex (1871), which extends the theory of natural selection to human evolution.

Darwin must be seen as a great intellectual revolutionary who inaugurated a new era in the cultural history of humankind, an era that was the second and final stage of the Copernican revolution that had begun in the 16th and 17th centuries under the leadership of men such as Nicolaus Copernicus, Galileo, and Isaac Newton. The Copernican revolution marked the beginnings of modern science. Discoveries in astronomy and physics overturned traditional conceptions of the universe. Earth no longer was seen as the centre of the universe but was seen as a small planet revolving around one of myriad stars; the seasons and the rains that make crops grow, as well as destructive storms and other vagaries of weather, became understood as aspects of natural processes; the revolutions of the planets were now explained by simple laws that also accounted for the motion of projectiles on Earth.

The significance of these and other discoveries was that they led to a conception of the universe as a system of matter in motion governed by laws of nature. The workings of the universe no longer needed to be attributed to the ineffable will of a divine Creator; rather, they were brought into the realm of science—an explanation of phenomena through natural laws. Physical phenomena such as tides, eclipses, and positions of the planets could now be predicted whenever the causes were adequately known. Darwin accumulated evidence showing that evolution had occurred, that diverse organisms share common ancestors, and that living beings have changed drastically over the course of Earth's history. More important, however, he extended to the living world the idea of nature as a system of matter in motion governed by natural laws.

Before Darwin, the origin of Earth's living things, with their marvelous contrivances for adaptation, had been attributed to the design of an omniscient God. He had created the fish in the waters, the birds in the air, and all sorts of animals and plants on the land. God had endowed these creatures with gills for breathing, wings for flying, and eyes for seeing, and he had coloured birds and flowers so that human beings could enjoy them and recognize God's wisdom. Christian theologians, from Aquinas on, had argued that the presence of design, so evident in living beings, demonstrates the existence of a supreme Creator; the argument from design was Aquinas's "fifth way" for proving the existence of God. In 19th-century England the eight Bridgewater Treatises were commissioned so that eminent scientists and philosophers would expand on the marvels of the natural world and thereby set forth "the Power , wisdom, and goodness of God as manifested in the Creation."

The British theologian William Paley in his Natural Theology (1802) used natural history, physiology, and other contemporary knowledge to elaborate the argument from design. If a person should find a watch, even in an uninhabited desert, Paley contended, the harmony of its many parts would force him to conclude that it had been created by a skilled watchmaker; and, Paley went on, how much more intricate and perfect in design is the human eye, with its transparent lens, its retina placed at the precise distance for forming a distinct image, and its large nerve transmitting signals to the brain.

The argument from design seems to be forceful. A ladder is made for climbing, a knife for cutting, and a watch for telling time; their functional design leads to the conclusion that they have been fashioned by a carpenter, a smith, or a watchmaker. Similarly, the obvious functional design of animals and plants seems to denote the work of a Creator. It was Darwin's genius that he provided a natural explanation for the organization and functional design of living beings. (For additional discussion of the argument from design and its revival in the 1990s, see below Intelligent design and its critics.)

Darwin accepted the facts of adaptation—hands are for grasping, eyes for seeing, lungs for breathing. But he showed that the multiplicity of plants and animals, with their exquisite and varied adaptations, could be explained by a process of natural selection, without recourse to a Creator or any designer agent. This achievement would prove to have intellectual and cultural implications more profound and lasting than his multipronged evidence that convinced contemporaries of the fact of evolution.

Darwin's theory of natural selection is summarized in the Origin of Species as follows:

As many more individuals are produced than can possibly survive, there must in every case be a struggle for existence, either one individual with another of the same species, or with the individuals of distinct species, or with the physical conditions of life....Can it, then, be thought improbable, seeing that variations useful to man have undoubtedly occurred, that other variations useful in some way to each being in the great and complex battle of life, should sometimes occur in the course of thousands of generations? If such do occur, can we doubt (remembering that many more individuals are born than can possibly survive) that individuals having any advantage, however slight, over others, would have the best chance of surviving and of procreating their kind? On the other hand, we may feel sure that any variation in the least degree injurious would be rigidly destroyed. This preservation of favourable variations and the rejection of injurious variations, I call Natural Selection.

Natural selection was proposed by Darwin primarily to account for the adaptive organization of living beings; it is a process that promotes or maintains adaptation. Evolutionary change through time and evolutionary diversification (multiplication of species) are not directly promoted by natural selection, but they often ensue as by-products of natural selection as it fosters adaptation to different environments.

MODERN CONCEPTIONS

The Darwinian aftermath

The publication of the Origin of Species produced considerable public excitement. Scientists, politicians, clergymen, and notables of all kinds read and discussed the book, defending or deriding Darwin's ideas. The most visible actor in the controversies immediately following publication was the English biologist T.H. Huxley, known as "Darwin's bulldog," who defended the theory of evolution with articulate and sometimes mordant words on public occasions as well as in numerous writings. Evolution by natural selection was indeed a favourite topic in society salons during the 1860s and beyond. But serious scientific controversies also arose, first in Britain and then on the Continent and in the United States.

One occasional participant in the discussion was the British naturalist Alfred Russel Wallace, who had hit upon the idea of natural selection independently and had sent a short manuscript about it to Darwin from the Malay Archipelago, where he was collecting specimens and writing. On

July 1, 1858, one year before the publication of the Origin, a paper jointly authored by Wallace and Darwin was presented, in the absence of both, to the Linnean Society in London—with apparently little notice. Greater credit is duly given to Darwin than to Wallace for the idea of evolution by natural selection; Darwin developed the theory in considerably more detail, provided far more evidence for it, and was primarily responsible for its acceptance. Wallace's views differed from Darwin's in several ways, most importantly in that Wallace did not think natural selection sufficient to account for the origin of human beings, which in his view required direct divine intervention.

A younger English contemporary of Darwin, with considerable influence during the latter part of the 19th and in the early 20th century, was Herbert Spencer. A philosopher rather than a biologist, he became an energetic proponent of evolutionary ideas, popularized a number of slogans, such as "survival of the fittest" (which was taken up by Darwin in later editions of the Origin), and engaged in social and metaphysical speculations. His ideas considerably damaged proper understanding and acceptance of the theory of evolution by natural selection. Darwin wrote of Spencer's speculations:

His deductive manner of treating any subject is wholly opposed to my frame of mind....His fundamental generalizations (which have been compared in importance by

some persons with Newton's laws!) which I dare say may be very valuable under a philosophical point of view, are of such a nature that they do not seem to me to be of any strictly scientific use.

Most pernicious was the crude extension by Spencer and others of the notion of the "struggle for existence" to human economic and social life that became known as social Darwinism (see below Scientific acceptance and extension to other disciplines).

The most serious difficulty facing Darwin's evolutionary theory was the lack of an adequate theory of inheritance that would account for the preservation through the generations of the variations on which natural selection was supposed to act. Contemporary theories of "blending inheritance" proposed that offspring merely struck an average between the characteristics of their parents. But as Darwin became aware, blending inheritance (including his own theory of "pangenesis," in which each organ and tissue of an organism throws off tiny contributions of itself that are collected in the sex organs and determine the configuration of the offspring) could not account for the conservation of variations, because differences between variant offspring would be halved each generation, rapidly reducing the original variation to the average of the preexisting characteristics.

The missing link in Darwin's argument was provided by Mendelian genetics. About the time the Origin of Species was published, the Augustinian monk Gregor Mendel was starting a long series of experiments with peas in the garden of his monastery in Brünn, Austria-Hungary (now Brno, Czech Republic). These experiments and the analysis of their results are by any standard an example of masterly scientific method. Mendel's paper, published in 1866 in the Proceedingsof the Natural Science Society of Brünn, formulated the fundamental principles of the theory of heredity that is still current. His theory accounts for biological inheritance through particulate factors (now known as genes) inherited one from each parent, which do not mix or blend but segregate in the formation of the sex cells, or gametes.

Mendel's discoveries remained unknown to Darwin, however, and, indeed, they did not become generally known until 1900, when they were simultaneously rediscovered by a number of scientists on the Continent. In the meantime, Darwinism in the latter part of the 19th century faced an alternative evolutionary theory known as neo-Lamarckism. This hypothesis shared with Lamarck's the importance of use and disuse in the development and obliteration of organs, and it added the notion that the environment acts directly on organic structures, which explained their adaptation to the way of life and environment of the organism. Adherents of this theory discarded natural selection as an explanation for adaptation to the environment.

Prominent among the defenders of natural selection was the German biologist August Weismann, who in the 1880s published his germ plasm theory. He distinguished two substances that make up an organism: the soma, which comprises most body parts and organs, and the germ plasm, which contains the cells that give rise to the gametes and hence to progeny. Early in the development of an egg, the germ plasm becomes segregated from the somatic cells that give rise to the rest of the body. This notion of a radical separation between germ plasm and soma—that is, between the reproductive tissues and all other body tissues—prompted Weismann to assert that inheritance of acquired characteristics was impossible, and it opened the way for his championship of natural selection as the only major process that would account for biological evolution. Weismann's ideas became known after 1896 as neo-Darwinism.

The synthetic theory

The rediscovery in 1900 of Mendel's theory of heredity, by the Dutch botanist and geneticist Hugo de Vries and others, led to an emphasis on the role of heredity in evolution. De Vries proposed a new theory of evolution known as mutationism, which essentially did away with natural selection as a major evolutionary process. According to de Vries (who was joined by other geneticists such as William Bateson in England), two kinds of variation take place in organisms. One is the "ordinary" variability observed among individuals of a species, which is of no lasting consequence in evolution because, according to de Vries, it could not "lead to a transgression of the species border [i.e., to establishment of new species] even under conditions of the most stringent and continued selection." The other consists of the changes brought about by mutations, spontaneous alterations of genes that result in large modifications of the organism and give rise to new species: "The new species thus originates suddenly , it is produced by the existing one without any visible preparation and without transition."

Mutationism was opposed by many naturalists and in particular by the so-called biometricians, led by the English statistician Karl Pearson, who defended Darwinian natural selection as the major cause of evolution through the cumulative effects of small, continuous, individual variations (which the biometricians assumed passed from one generation to the next without being limited by Mendel's laws of inheritance [see Mendelism]).

The controversy between mutationists (also referred to at the time as Mendelians) and biometricians approached a resolution in the 1920s and '30s through the theoretical work of geneticists. These scientists used mathematical arguments to show, first, that continuous variation (in such characteristics as body size, number of eggs laid, and the like) could be explained by Mendel's laws and, second, that natural selection acting cumulatively on small variations could yield major evolutionary changes in form and function. Distinguished members of this group of theoretical geneticists were R.A. Fisher and J.B.S. Haldane in Britain and Sewall Wright in the United States. Their work contributed to the downfall of mutationism and, most important, provided a theoretical framework for the integration of genetics into Darwin's theory of natural selection. Yet their work had a limited impact on contemporary biologists for several reasons—it was formulated in a mathematical language that most biologists could not understand; it was almost exclusively theoretical, with little empirical corroboration; and it was limited in scope, largely omitting many issues, such as speciation (the process by which new species are formed), that were of great importance to evolutionists.

A major breakthrough came in 1937 with the publication of Genetics and the Origin of Species by Theodosius Dobzhansky, a Russian-born American naturalist and experimental geneticist. Dobzhansky's book advanced a reasonably comprehensive account of the evolutionary process in genetic terms, laced with experimental evidence supporting the theoretical argument.

Genetics and the Origin of Species may be considered the most important landmark in the formulation of what came to be known as the synthetic theory of evolution, effectively combining Darwinian natural selection and Mendelian genetics. It had an enormous impact on naturalists and experimental biologists, who rapidly embraced the new understanding of the evolutionary process as one of genetic change in populations. Interest in evolutionary studies was greatly stimulated, and contributions to the theory soon began to follow, extending the synthesis of genetics and natural selection to a variety of biological fields.

The main writers who, together with Dobzhansky, may be considered the architects of the synthetic theory were the German-born American zoologist Ernst Mayr, the English zoologist Julian Huxley, the American paleontologist George Gaylord Simpson, and the American botanist George Ledyard Stebbins. These researchers contributed to a burst of evolutionary studies in the traditional biological disciplines and in some emerging ones—notably population genetics and, later, evolutionary ecology (see community ecology). By 1950 acceptance of Darwin's theory of evolution by natural selection was universal among biologists, and the synthetic theory had become widely adopted.

Molecular biology and Earth sciences

The most important line of investigation after 1950 was the application of molecular biology to evolutionary studies. In 1953 the American geneticist James Watson and the British biophysicist Francis Crick deduced the molecular structure of DNA (deoxyribonucleic acid), the hereditary material contained in the chromosomes of every cell's nucleus. The genetic information is encoded within the sequence of nucleotides that make up the chainlike DNA molecules. This information determines the sequence of amino acid building blocks of protein molecules, which include, among others, structural proteins such as collagen, respiratory proteins such as hemoglobin, and numerous enzymes responsible for the organism's fundamental life processes. Genetic information contained in the DNA can thus be investigated by examining the sequences of amino acids in the proteins.

In the mid-1960s laboratory techniques such as electrophoresis and selective assay of enzymes became available for the rapid and inexpensive study of differences among enzymes and other proteins. The application of these techniques to evolutionary problems made possible the pursuit of issues that earlier could not be investigated—for example, exploring the extent of genetic variation in natural populations (which sets bounds on their evolutionary potential) and determining the amount of genetic change that occurs during the formation of new species.

Comparisons of the amino acid sequences of corresponding proteins in different species provided quantitatively precise measures of the divergence among species evolved from common ancestors, a considerable improvement over the typically qualitative evaluations obtained by comparative anatomy and other evolutionary subdisciplines. In 1968 the Japanese geneticist Motoo Kimura proposed the neutrality theory of molecular evolution, which assumes that, at the level of the sequences of nucleotides in DNA and of amino acids in proteins, many changes are adaptively neutral; they have little or no effect on the molecule's function and thus on an organism's fitness within its environment. If the neutrality theory is correct, there should be a "molecular clock" of evolution; that is, the degree to which amino acid or nucleotide sequences diverge between species should provide a reliable estimate of the time since the species diverged. This would make it possible to reconstruct an evolutionary history that would reveal the order of branching of different lineages, such as those leading to humans, chimpanzees, and orangutans, as well as the time in the past when the lineages split from one another. During the 1970s and '80s it gradually became clear that the molecular clock is not exact; nevertheless, into the early 21st century it continued to provide the most reliable evidence for reconstructing evolutionary history. (See below The molecular clock of evolution and The neutrality theory of molecular evolution.)

The laboratory techniques of DNA cloning and sequencing have provided a new and powerful means of investigating evolution at the molecular level. The fruits of this technology began to accumulate during the 1980s following the development of automated DNA-sequencing machines and the invention of the polymerase chain reaction (PCR), a simple and inexpensive technique that obtains, in a few hours, billions or trillions of copies of a specific DNA sequence or gene. Major research efforts such as the Human Genome Projectfurther improved the technology for obtaining long DNA sequences rapidly and inexpensively. By the first few years of the 21st century, the full DNA sequence—i.e., the full genetic complement, or genome—had been obtained for more than 20 higher organisms, including human beings, the house mouse (Mus musculus), the rat Rattus norvegicus, the vinegar fly Drosophila melanogaster, the mosquito Anopheles gambiae, the nematode worm Caenorhabditis elegans, the malaria parasite Plasmodium falciparum, the mustard weed Arabidopsis thaliana, and the yeast Saccharomyces cerevisiae, as well as for numerous microorganisms. Additional research during this time explored alternative mechanisms of inheritance, including epigeneticmodification (the chemical modification of specific genes or gene-associated proteins), that could explain an organism's ability to transmit traits developed during its lifetime to its offspring.

The Earth sciences also experienced, in the second half of the 20th century, a conceptual revolution with considerable consequence to the study of evolution. The theory of plate tectonics, which was formulated in the late 1960s, revealed that the configuration and position of the continents and oceans are dynamic, rather than static, features of Earth. Oceans grow and shrink, while continents break into fragments or coalesce into larger masses. The continents move across Earth's surface at rates of a few centimetres a year, and over millions of years of geologic history this movement profoundly alters the face of the planet, causing major climatic changes along the way. These previously unsuspected massive modifications of Earth's past environments are, of necessity, reflected in the evolutionary history of life. Biogeography, the evolutionary study of plant and animal distribution, has been revolutionized by theknowledge, for example, that Africa and South America were part of a single landmass some 200 million years ago and that the Indian subcontinent was not connected with Asia until geologically recent times.

Ecology, the study of the interactions of organisms with their environments, has evolved from descriptive studies—"natural history"—into a vigorous biological discipline with a strong mathematical component, both in the development of theoretical models and in the collection and analysis of quantitative data. Evolutionary ecology (seecommunity ecology) is an active field of evolutionary biology; another is evolutionary ethology, the study of the evolution of animal behaviour. Sociobiology, the evolutionary study of social behaviour, is perhaps the most active subfield of ethology. It is also the most controversial, because of its extension to human societies.

THE CULTURAL IMPACT OF EVOLUTIONARY THEORY

Scientific acceptance and extension to other disciplines

The theory of evolution makes statements about three different, though related, issues: (1) the fact of evolution—that is, that organisms are related by common descent; (2) evolutionary history—the details of when lineages split from one another and of the changes that occurred in each lineage; and (3) the mechanisms or processes by which evolutionary change occurs.

The first issue is the most fundamental and the one established with utmost certainty. Darwin gathered much evidence in its support, but evidence has accumulated continuously ever since, derived from all biological disciplines. The evolutionary origin of organisms is today a scientific conclusion established with the kind of certainty attributable to such scientific concepts as the roundness of Earth, the motions of the planets, and the molecular composition of matter. This degree of certainty beyond reasonable doubt is what is implied when biologists say that evolution is a "fact"; the evolutionary origin of organisms is accepted by virtually every biologist.

But the theory of evolution goes far beyond the general affirmation that organisms evolve. The second and third issues—seeking to ascertain evolutionary relationships between particular organisms and the events of evolutionary history, as well as to explain how and why evolution takes place—are matters of active scientific investigation. Some conclusions are well established. One, for example, is that the chimpanzee and the gorilla are more closely related to humans than is any of those three species to the baboon or other monkeys. Another conclusion is that natural selection, the process postulated by Darwin, explains the configuration of such adaptive features as the human eye and the wings of birds. Many matters are less certain, others are conjectural, and still others—such as the characteristics of the first living things and when they came about—remain completely unknown.

Since Darwin, the theory of evolution has gradually extended its influence to other biological disciplines, from physiology to ecology and from biochemistry to systematics. All biological knowledge now includes the phenomenon of evolution. In the words of Theodosius Dobzhansky, "Nothing in biology makes sense except in the light of evolution."

The term evolution and the general concept of change through time also have penetrated into scientific language well beyond biology and even into common language. Astrophysicists speak of the evolution of the solar system or of the universe; geologists, of the evolution of Earth's interior; psychologists, of the evolution of the mind; anthropologists, of the evolution of cultures; art historians, of the evolution of architectural styles; and couturiers, of the evolution of fashion. These and other disciplines use the word with only the slightest commonality of meaning—the notion of gradual, and perhaps directional, change over the course of time.

Toward the end of the 20th century, specific concepts and processes borrowed from biological evolution and living systems were incorporated into computational research, beginning with the work of the American mathematician John Holland and others. One outcome of this endeavour was the development of methods for automatically generating computer-based systems that are proficient at given tasks. These systems have a wide variety of potential uses, such as solving practical computational problems, providing machines with the ability to learn from experience, and modeling processes in fields as diverse as ecology, immunology, economics, and even biological evolution itself.

5

Environmental Management System

Rapid movement of industrialization throughout the world has been seriously menacing mankind's ability to maintain an ecological balance. Industrialization is the prerequisite of economic growth of any country; but unplanned industrialization and release of waste by industries brings environmental pollution or degradation. Every developmental activity directly relates to natural and environmental resources. Economic development without environmental considerations causes to environmental crises. The ecological role played by corporate sectors responsible for their business activities on the environment is becoming particularly explicit in the global market, especially in India. Management of both environmental and natural resources in a country like India has become more urgent. Every industry feels the necessity of sustainable development management. In addition, global awareness and acceptance of the importance of environmental issues has motivated the development of a new area of management known as "Corporate Environmental Management". Environmental management system is a part of the overall management system. The destructive influence of man's economic activities on the environment in the name of sustainable global economy is not of recent origin.

In India, government has come forward to exercise control over the issue relating to degradation of environment. It is worth mentioning in this respect that the Indian constitution imposes as one of the fundamental duties of every citizen, the duty to protect and improve the natural environment including forests, lakes, rivers and wild life and to have compassion for living creatures.

ENVIRONMENTAL MANAGEMENT – CONCEPT

Environmental Management is entirely an emerging and dynamic concept. Environmental Management is concerned with the management for environment encompassing a business. It represents the organizational structure, responsibilities sequences, processes and preconditions for the implementation of an environmental corporate policy. Environment brings together all inanimate organism and forces functioning in nature including man. The basic functions of good environmental management are goal setting; information management; support of decision making; organizing and planning of environmental management; environmental management programs; piloting; implementation and control; communication; internal and external auditing, etc. The present state of economic development, including the environmental state, makes it necessary to broaden management's understanding of natural environment. The way of industrialization being emphasized for the development of economy, in coming year's environmental pollution will be the ecological nightmare. Hence, it has become imperative to take into account the ecological consequences while setting up an industrial unit. Technology is available today to reduce the environmental pollution and it must be used to correct the excesses of ecological brutality and to minimize the degree of environmental degradation. For all these, a proper accounting and reporting of environmental information is a must which can lead to sound "Environmental Management".

Concept of Environmental Management:

The ecological balance and ecosystem stability are duly maintained by the nature itself but the emergence of modern industrial era has disturbed the ecological balance through heavy industrialization, technological revolution, faster growth of means of transportation, rapacious exploitation of resources, unplanned urbanization etc..

In other words, the anthropogenic activities of modern 'economic and technological' man have disturbed the harmonious relationships between the environment and human beings. Environmental management is thus, the process to improve the relationship between the human beings and environment which may be achieved through check on destructive activities of man, conservation, protection, regulation and regeneration of nature.

The process, environmental management is related to the rational adjustment of man with nature involving judicious exploitation and utilization of natural resources without disturbing the ecosystem balance and ecosystem equilibrium.

If the natural resources are overexploited, it will affect socio-economic development of a nation. Thus, environmental management must take into consideration the ecological principles and socioeconomic needs of the society i.e., it involves socio economic developments on one hand and maintenance of environmental quality on other hand.

From the above discussion, it is clear that environmental managements has two major aspects:

(i) ocio-economic development and

(ii) Stability of biosphere in general and stability of individual ecosystems in particular (C.C. Park 1981).

Scope and Aspects of Environmental Management:

Environmental management is very wide in scope and includes all the technical, economical and other aspects of environment.

The broader objectives of environmental management includes:

(i) To identify the environmental problem and to find its solution.

(ii) To restrict and regulate the exploitation and utilization of natural resources.

(iii) To regenerate degraded environment and to renew natural resources (renewable)

(iv) To control environmental pollution and gradation.

(v) To reduce the impacts of extreme events and natural disaster.

(vi) To make optimum utilization of natural resources.

(vii) To assess the impacts of proposed projects and activities on environment.

(viii) To review and revise the existing technologies and make them ecofriendly.

(ix) To formulate laws for the implementation of environmental protection and conservation programmes.

The components of environmental management are based on five fundamental aspects.

1. Environmental perception and public awareness:

The environmental perception and public awareness considers the following points:

(a) Sources of environmental perception and public awareness.

(b) Level of environmental perception.

(c) Role of environmental perception in environmental planning and management.

2. Environmental education and training:

Environmental education and training should be given at school, college and University levels by professionals.

3. Resource management:

The resource management considers the following points:

(i) Classification of natural resources

   (ii) Survey and evaluation of ecological resources

   (iii) Preservation of resources

   (iv) Conservation of resources

4. Control of Environmental degradation and pollution:

The environmental degradation and pollution can be checked by considering the following points:

   (i) Control of environmental degradation and pollution.

   (ii) Adopting suitable preventive mechanisms to reduce natural hazards and disaster.

   (iii) Regeneration of degraded environment.

5. Environmental impact assessment:

The environmental impact assessment involves:

   (i) ppraisal of existing environmental conditions

   (ii) Appraisal of existing and proposed production methods

   (iii) Methologies and procedures

   (iv) Probable impacts of existing and proposed project.

   (v) Review of technology and required improvement.

RESEARCH OBJECTIVE AND METHODOLOGY

This paper being basically based on concepts, the opinions expressed in this paper are the author's own opinion and the opinions of some reputed authors. This paper shows light on certain fundamental and theoretical aspects of the concept of "Environmental Management". The study also attempts to examine the importance of environmental management today. In the light of this, the motivation for this paper is not to seek new solution to the underlying problems facing environmental management research. This discussion is expected to provide insights and a basis from which management can test their current practices and seek to encourage them in terms of providing a better understanding of the interaction between corporate and the natural environment. Special importance is placed on environmental management and awareness since this is supposedly the crying need of the day.

Importance of Environmental Management

Environment takes into consideration all conditions required for the survival of corporate sectors. Absence of environmental consideration causes serious ecological damage. Poverty, lack of resources, population pressure and global inequity of the resource use are generating unparalleled social and environmental problems at national and global levels. Sustainable development has a tendency to strike a balance between the demands of economic development and the need for management of environmental processes. Precise knowledge of various facets of sound environmental management is the sine qua non for sustainable development which meets not only the requirements of the present generation but also of the future generation. Environmental management is especially valuable for internal management initiatives with a specific environmental focus, such as cleaner production, supply chain management, "given" product or service design, environmentally preferable purchasing and environmental management systems. Environmental management type information is increasingly being used for external reporting purposes also. Environmental Management Accounting (EMA) is one of the environmental management tools. EMA provides a broad set of principles and approach required for the success of many other environmental management functions and since the range of decisions affected by environmental issues is gradually increasing, EMA is becoming more important not only for environmental management decisions, but for all types of managerial functions with special emphasis on eco-efficiency and strategic position.

As a result, environmental management helps corporate and other organizations boost their public trust and confidence and are related to receiving a fair assessment. The already degraded environment calls for its diligent management. Through environmental management, enterprises can enjoy the following benefits:

i. Pollution control being burning subject of discussion, environmental management shows the extent to which pollution has been controlled by the corporate.

ii. Environmental management draws attention in another sense, which offers an idea about industrial development, a nation's economic progress and social welfare and the fulfillment of responsibility towards society.

iii. Environmental management is helpful in discharging organizational accountability and increasing environmental transparency. Sustainable development is possible with the help of environmental management as it helps include ecological ability of enterprise.

iv. Negotiation between the management and society helps organizations seek to strategically mange a fresh and emerging issue with distinct users.

v. Environmental management supports green reporting to combat effectively all negative public opinions in the global economy where existence of a strong environmental lobby against environmentally unfriendly industries is found.

vi. Environmental management improves performance through better management of environmental cost and thus, benefits the natural and human environments.

vii. Environmental management forces corporate sectors to fulfill their commitments towards introduction and change, and thus appears to be responsive to new factors. Countries giving importance to the ecological aspect of activities are becoming more and more popular, particularly in Western Countries.

viii. Environmental management reflects unsound production and consumption patterns, misuse and scanty use of resources and assets like water.

ix. Optimal allocation of scanty resources in the economy is possible with the help of environmental management.

x. Environmental management is essential in measuring a nation's economic development, social welfare, industrial development, pollution control and in fulfilling the needs of government, still the system is in its infancy and not all countries have been able to develop such a system [1]. But with the passage of time, the system will gradually develop, research will be undertaken and it will fulfill the requirements for which it was originated.

xi. Impressive decision taking through the application of environmental management

reduces or eliminates many environmental problems.

In the light of evolutionary learning, it is considered essential to make an endeavor to incorporate the effects of environmental resources in the entire business functions of a business house. Environmental management through environmental accounting is an attempt to identify the resources consumed and the costs imposed on the environment by a corporate. Every corporate citizen should evaluate an environmental management system through environmental accounting to keep records of the benefits and costs rendered by the environment of a corporate and justifying these costs and benefits are large constituents of environmental accounting which in turn, nourishes a corporate development and operation for an all-round environmental management system. Maintaining secrecy and widely following different accounting systems sometimes make the task very difficult. But still, environmental management is highly useful in planning, public resource management, pollution control and for policy analysis.

Corporate Agenda

The responsibilities of accountants in environmental management cannot be denied. An accountant should create environmental consciousness in the corporate sectors from top level to bottom level so that the environmental culture may be installed and their cooperation may be sought for in enforcing it. Every staff in the corporate sector must be made aware of the importance of environmental quality management. The management accountant has an important role in uttering this principle or mantra in every walk of corporate life. Corporate agenda should be prepared and the following may be incorporated there:

    i. Corporate commitment to environmental protection should be clearly stated;

    ii. Environmental protection is a genuine task and no stone will be left unturned in preserving and safeguarding ecological balances;

    iii. Environmental friendly code of conduct must be noticed at all levels and by all the people;

    iv. Employees failing to observe environmental friendly code of conduct must be reprimanded;

    v. Alteration in product design should be made as far as possible to incorporate eco-friendly materials and other inputs;

    vi. Renewable sources of energies like solar energy, wind energy, etc. must be used and so advocated;

    vii. Maintenance, replacement and repairs of machinery, tools, equipment, vehicles, etc. should be done properly and in time;

    viii. Arrangement for returning or collecting back non-biodegradable packing materials must be used to the best possible extent;

    ix. If the working of the corporate on any day or in any month is against environmental interest, staff doing this must seek apology;

    x. Installation of adequate waste and sewage plants before the industrial wastages are left into river, sea, etc. may be made;

    xi. Adequate number of PROs to communicate with the external world and to clear the corporate commitment to the world at large must be recruited;

    xii. Environmental engineers must be recruited for seeking their opinion on environmental related matters and for involvement in establishing a cleaner and efficient production system thereby minimizing wastage and ensuring optimum usage of all the raw materials;

    xiii. Corporate manager should be proactive rather than active. The adequate measures must be initiated to protect environment before the new technology is put to use or adopted;

    xiv. Participation in activities like social forestry, community forest efforts, environment cleaning works, etc. may be made;

xv. R&D activities for identifying new processes or methods of production or inputs which are sustainable from environmental point of view should be encouraged. The aforesaid corporate environmental agenda should be drafted by the management accountant and implemented in and outside the corporate with full support of the top management. The management accountant as far as environmental matters are concerned can be taken as the "spokesperson" who represents the environment. Regular environmental performance reports, environmental laws and changes in them, periodical environmental audit by independent agencies, etc. should be brought to the notice of all concerned in the corporate to ensure genuine environment to protect the environment.

Importance of Environmental Balance Sheet

Environmental balance sheet may be prepared and attached along with the annual financial statement. Environmental balance sheet summarizes a corporate environmental performance and its net environmental capital investment in the shape of environment-friendly assets, its environmental liabilities and its net environmental capital. It is always desirable from environmental protection point of view for a firm to possess positive contribution. Environmental balance sheet usually includes the following:

Environmental Assets: Water pollution treatment plants; Sewage treatment plants; Air pollution preventive systems; Investment in Social forestry; Human Training and Development cost; Capital work-in-progress; Renewable energy sources; Eco-friendly production machines; Packaging materials etc. The above assets are required to control environmental degradation.

Environmental Liabilities: Amount of compensation payable according to court rulings; NGOs' rulings; Pending Law suits against environmental hazards; Amount of damages yet to be assessed, etc. Environmental liabilities are actually the obligations payable or incurred in discharge of environmental responsibilities by a corporate. These liabilities can highly influence the corporate financial status. A positive financial net worth could be converted into negative one if the working of the corporate is at the cost of environmental protection. It is, therefore, essential to assess them properly in determining environmental capital. The objective of any corporate is to minimize its environmental liabilities by being friendly to its environment. Green capital takes place when environmental investments exceed the environmental liabilities.

It is desirable that environmental capital should be equal to environmental liabilities. Environmental balance sheet is an indispensable part of yearly financial accounts. Necessary legal provisions may be made to make it mandatory for the corporate. Further, Courts, NGOs, Consumer Forums, Environmental and Social workers and other environmental organizations should strive for necessary morale persuasion on the corporate management to prepare and publish the amount of Green Capital investment in environmental protection.

Necessity of Accounting Standard

Environmental management through environmental accounting is the crying need of the hour. Hence, with a view to measuring and reporting of environmental investment, an accounting standard is of utmost importance. Measurement of environmental hazards created by corporate is a very difficult task and also the amount of damage differs from concern to concern depending on the nature of industry, nature of management, its philosophy, methods of production and distribution, environmental pollution, motivation level of employees, reward and reprimand policies etc. But the standard to be introduced must dovetail the environmental hazards created by different industries and the cost incurred by the corporate and by the society to alleviate bad consequences. Such type of accounting standard setting at national and international levels must include environmental scientists, representatives of NGOs, Government agencies, Corporate, UN representatives and above all accountants. Standard practices of reporting must also be

Benefits of Environmental Management Accounting

Environmental Management Accounting (EMA) has no single univocally accepted definition. EMA is the management of environmental and economic performance with the help of development and execution of appropriate environmentally related accounting system and practices. EMA typically includes reporting, auditing, life-cycle costing, full-cost accounting benefits assessment and strategic planning for environmental management. EMA identifies collects, analyses and uses of two types of information for internal decision making purpose viz. 1. Physical information on the use, flows and destinies of energy, water and materials (including wastes) and 2. Financial information on environment related costs, earnings and savings. Although EMA brings light basically on internal management decision making, physical accounting information also is often reported to external stakeholders.

ENVIRONMENTAL AWARENESS IN INDIA

Problems

Environmental management through environmental accounting shows the extent of pollution controlled by business entities. Man has been rapidly and deliberately exploiting the environmental resources with the aid of modern science and technology. Industrialization is genuine for life, but evils accompanying it are also no less in number. The most outstanding and patent danger that emerges from the industrial activities is pollution. In underdeveloped countries, pollution is not the serious problem as it is in technologically developed countries of the world. In well-developed nations and the greatest technologically advanced countries, the worst pollution happens. The industrialization is mainly concerned with physical environmental pollution (i.e. air, water and noise). Most of the Indian rivers and fresh water streams are seriously polluted by industrial wastes or elements of different industries causing waterborne diseases. Unplanned urbanization, construction of water projects, and migration of people – everything helps change the ecology and epidemiology of diseases. India is the third largest producer of tobacco in the world after the US and China. The Government of India has done little to control or reduce smoking because of conflicting loyalties, the need of the exchequer and the health of the people. Today unfortunately, urban ecology is no sounder and is poised with health hazards and impaired human activity, due to low per capita availability of land. Environment and animal are both polluting each other. Untreated hospital wastes in the garbage places are endangering the health of both animal and man. Drossy animals on the urban roads are exposed to serious chemical pollution from automobile exhausts that lour their health, productivity and also reproductive efficiency. Environmental pollution in India is a serious problem now and serious efforts are being made to orient the public in its protection.

The administration of pollution control has been gloomy due to want of reliable information on the violation and implementation of these regulations. In India, we are not better off. The worst polluting company appears to be the most successful and attracts additional investment from an investing public. Measuring the inter-relationship between environment and development has become a complicated issue mainly because the current national accounting does not only consider the value of natural resources. The problem of environmental protection is becoming more and more acute and the necessity for considering the value of services of environmental resources is also gradually increasing day by day. Monetary values allotted to environmental goods and services under the shadow pricing process are not certain and inadequately quantified. Many conditions and assumptions lying under economic theory are not encountered. International firms and organizations had a tendency to disclose non-monetary information including environmental information; naturally, they have enhanced the expectations from Indian corporate to act responsively towards the environment and be accountable to society beyond the traditional role of providing financial accounts to shareholders. Usually, literature available as to environmental performance reporting has concentrated on developed countries and little attention has been given to the states of environmental reporting of developing countries. So far, no accounting standard has been issued extensively for accounting treatment of the aforesaid problems. Though some guidelines in regard to these problems have been issued by many organizations internationally, but these guidelines are mostly advisory in nature. A large number of countries impose requirements on corporate to report on their environmental performance, in India, corporate are required to prepare a so-called "Green Account". Absence of comprehensive and verifiable information and financial data on environmental performance of companies sometimes induce them to pollute the environmental and yet appear more efficient economically than others which incur costs to protect the environment. The economics approach to environment issues as reflected in some countries opines that companies in unscrupulous pursuit of profits can do much social harm and the environment suffers. Environmental management is not effectively followed in India. Environmental aspects in India are so significant to a company that there is every possibility or risk for the occurrence of significant misrepresentation or inadequate or incomplete presentation of such information within financial statements.

Awareness

Laws for the preservation of environment were passed in India in 1970. Since then, environmental awareness among the public in India has grown tremendously. Smt. Indira Gandhi, the then Prime Minister of India felt the necessity of healthy and clean environment and expressed her energetic opinion in favor of the environment at the United Nation's Conference on the human environment in Stockholm, 1972. Diverse rules, laws, etc. have been enacted from time to time and India has been facing strong competition particularly after the opening of the Indian economy in 1990. The first public announcement in regard to environmentally related information from business on a periodic basis was made by the Central Government in 1991. Environmental management through keeping of accounts and records with the help of environmental accounting would be beneficial to hinder exploitation of natural resources and prevent their depletion. The Environmental Reporting Practices of the corporate sectors in India are not satisfactory; very few corporate entities have some mention about the environment in general terms. Most of the corporate entities have taken reporting on environmental aspects under more statutory obligations and less social responsibility. The statutory provisions about reporting on environment protection are not adequate to submit true and fair information about the consequences of corporate operation on environment. Although environmental management through environmental accounting and reporting is voluntary exercise in India, the organizations preferring to exhibit environmental issues in their statements enjoy several benefits such as improved image of the product or corporate.

Conclusion and Suggestion

Environment being an essential part of every body's living is said to be true even for legal entities i.e. corporate. The corporate living should not destroy environment balance sheet rather it should largely contribute for sustainable development of a country.

The social values placed on environmental goods and services are changing so rapidly that estimates are likely to be obsolete before they are available for use. Planning for sustainable development requires an estimate of environmentally adjusted GNP. However, despite the theoretical irregularities, the slogan for environmental management and environmental accounting has won perpetual benefit inherent in it. In addition, growing awareness and acceptance of the importance of natural and environmental resources globally and nationally has laid to the development of environmental management. Valuation of environmental goods and services and incorporation of environmental data into the national and corporate levels suggest different techniques. In many countries, the disclosure practices in regard to environmental issues have become mandatory. But in some countries, such mandate is not everywhere. Taking step internationally and particularly to formulate valuation techniques regarding environmental issues is now an urgent need. Mandatory guidelines can be issued in each and every country to incorporate these in the company's annual report including environmental related legislation as in developed countries [4]. The dedication with which work for the development of environmental management is going on will surely lead to environmental management occupying a more stable and efficacious position in the coming future, as it could greatly improve the value of economics as a decision-making tool, especially in determining national policy. The implementation of environmental management is expected to bring about a change in the managerial attitudes and thinking. Despite difficulties associated with environmental management, there is much evidence to show that a large number of countries around the world have sincerely attempted to pick-up the new challenges and threats. Economic activity should not be guided by 'profit motive' alone, but should also include "quality of life" and "ecological balance". The key to sustainable growth, therefore, is not to provide less but to provide efficiently with the help of environmental management system.

For developing harmonization following accounting policy, a comprehensive plan in connection with environmental management may be initiated for all types of companies. It creates an environmentally conscious new atmosphere in corporate sectors and prepares and publishes environmental balance sheets which would pave the way for the increased earnestness among corporate. Traditionally many internal environmental costs are not recognized, allocated or assigned to the activities. It is suggested that through environmental management, companies should fully recognize and control all environmental costs, including the aforesaid costs. A business should internalize these costs by anticipating and managing them. Environmental expenditures must be separated to improve decision making and accountability for environmental responsibilities. Every company should focus and set aside a part of their funds for meeting environmental and ecological balance. Unless comprehensive records for the use of natural resources and environment and their services is maintained, reliable and sustainable development cannot be expected. Therefore, it is considered essential to make an endeavor to incorporate the effect of environmental resources in the entire business function of a business corporation. The technology is available today to reduce environmental pollution and it must be used to correct the excess ecological brutality and minimize the degree of environmental pollution. Current disclosure practices by most of the corporate sectors in India do not entirely reflect the environmental impact of corporate performances. There is a need and challenge for companies to become greener when we are matching towards industrialization and globalization. Only an environmentally acceptable company has a secure future. Companies seeking for long-term profits should consider an ecological aspect in its business strategy and policy. The sooner precautions are taken to protect them, the better it would be. The already polluted environment calls for its persevering management. A large number of international agreements on environment have been negotiated having sustainable trade effects. In addition, some countries including India are taking unilateral steps directed towards environmental protection which also have trade implications for the partner countries. Undoubtedly, Indian industry has to go for higher level of environmental standards in future both because the country's environmental protection needs such standards as well as the demand of the important trade partners. The Government of India is spending a lot of money for removing water pollution, air pollution, noise pollution, etc. Central Board for the Prevention and Control of Water Pollution is making efforts for reducing and checking the water pollution throughout the country.

Thanks to our Government for establishing Central Ganga Authority (CGA) meant for cleansing of Ganga Water but proper utilization of funds is essential and this scheme should be extended to other rivers. In India air pollution control legislation envisages the formation of air pollution boards at the Central and State levels. Of late, many nations have enacted laws to penalize noise pollution by the vehicles or any type of industry. Government of India (State) has made constitutional provisions for environmental management. The Government of India will look after the environment not only on the instances of the courts and the specific provisions of the Indian Constitution, because mere legislation can never be sufficient if it is not enforced strictly. Our environmentalists should keep guard the enforcement of the anti-pollution laws so that profit-minded industrialists are compelled to adopt modern scientific techniques to minimize the pollution and noise pollution. "Back to nature" should be the slogan in the minds of people for protecting and survival of mankind and its health. The environmental management plan should have sufficient measures for minimizing the environmental damage, revival of mine areas and for planting of trees in accordance with the prescribed norms. Attempts should also be made to convert old mining sites into forests and other forms of land use. Recently, Government has set up labor Institutes across the country to deal with the problems of chemical safety. Love and kindness towards dumb animals is the inherent character of ancient Indian culture and they are considered sacred. The study and practices of bioethics should be made a part of veterinary profession and even in the primary education itself. Scientific and ethical disposal of industrial and domestic wastes has to be given due weight. Appropriate steps to isolate the sick and to maintain new animals under sufficient quarantine have to be followed to safeguard the cattle and human health. Educating everyone concerned directly and indirectly with the animals and their products and services needs special attention. It should be kept in mind that health education is a continuous process requiring intermittent re-orientation and follow up. Man should manifest positive attitudes towards animals and the animals will not be afraid of man, and there should be better social bound between man and animal, thus fulfilling the universal motto "Health and Wealth for All". Through several activities, NGOs definitely can subscribe to the overall development of villages which are the backbone of the country's economic progress. With clean aims and objectives, the NGOs are expected to play a vital role in the overall development of environment to benefit all sections of society and help generate a quality environment and nourish it for future generations. Environment friendly product is built on the non-exploitative treatment of natural resources. It is based on Gandhiji's advice, "Nature provides for everybody' s needs but not everyone's greed".

It is the spread of a greed revolution that causes harm to our life support systems. An ever-green evolution based on harnessing solar energy through green plants and adopting environment friendly agriculture practices can alone ensure opportunities for a productive and healthy life for all. It is hoped that the GRREN INDIA-2017 project would be a bold step in acquiring knowledge on these subjects in the right direction. It becomes necessary to develop a suitable program for creating environmental awareness among various sections of society and orient them for adopting environmental friendly production and consumption practices in day to day life. There is further scope for involvement of environmental education through well designed training courses, publications and documentary films to orient the children, educated people, industrial workers, farmers and others for 21 century. The pollution will not only pay for the damage done but also for restoration of the ecology destroyed by him/her in an area. Sadly, the official legal aid machinery has yet to wake up to the need of legally assisting million of the poor in slums and villages denied both the dignity and a healthy environment. Social and environmental audits are not yet very popular in India, however, they are in use for a long time in developed countries. The reason for their adoption lies in very serious public interest groups and social activists. In India, the demand for social audit and environmental audit is gradually increasing. The importance of audit and role of auditors is constantly increasing and this audit is now being adopted by many companies. The government has set up Central and State Pollution Control Boards and Laboratories and no industry can be set up without taking prior environmental clearance. Although developed countries have inadequate monitoring and regulating agencies. Of late, the courts are also taking a pro activist stand in cases related to environmental pollution. India is considered to be one of the few countries in the developing world which havecomprehensive regulations belonging to the environment. However, the administration of pollution control has been dreary due to want of reliable information on the violation and implementation of these regulations. In India, we, are not better off. We should not adopt Western complacency.

Rather, we should initiate an inter-disciplinary approach with the audit profession acting as the monitors and reporters by verifying, attesting and reporting on environmental disclosures to make an item a more reliable input for rewarding performers and penalizing polluters. Sound environmental management can only be equated with good management. Moreover, better environmental management ensures resource saving and helps cut down production cost. Recycle and reuse of wastes have led to cost saving in many chemical process industries. Several industries have adopted cleaner technologies that generate less waste and make production more profitable.

Industry can therefore, clearly benefit from a critical self-examination of the processes and technologies it employ to see in which areas there is scope for improvement and foresee the potential problem areas, particularly pollution and human health. Environmental accounting as a part of social and management accounting not only provides information to all, it also creates an environmental consciousness atmosphere in corporate and prepare and publish environmental balance sheets. An independent accounting standard on recording and reporting practices of environmental accounting would pave the way for the increased seriousness among corporate. The energy audit is key to a methodical approach for decision making and planning in the energy management program. Energy audit is an effective energy conservation tool to define and pursue a comprehensive energy management program in an industry. It is urgent need for all industries to introduce energy audit without fail and Government of India by regulations should also insist upon the same as compulsory for all industries. Further, annual energy audit has to be made mandatory for all industries to conserve the precious and non-re-perishable resources of our nation. A few other suggestions are:-

    i. A regular and periodic review and appraisal regarding the prevailing conditions and the improvements effected vis-à-vis the investments made and planned for .

    ii. Implementation of the plants with an eye to not only economic development of funds but also to selectivity of the contents and features of each of them.

    iii. Careful attention is necessary so that efforts at the national and corporate level of various bodies do not run counter to each other.

    iv. Creation of national parks for preservation of flora and fauna is an effort to protection of the environment from any kind of degradation.

    v. Development of the accounting formats for keeping records of all the efforts along with measurable and national advantages and disadvantages.

    vi. Annexure may be included in the Director's Report with necessary amendment in the Company's Act for development of pollution standards for industries or products vs. actual pollution generation position of the company.

6

Ecology

Ecology is the scientific analysis and study of interactions among organisms and their environment. It is an interdisciplinary field that includes biology, geography, and Earth science. Ecology includes the study of interactions that organisms have with each other, other organisms, and with abiotic components of their environment. Topics of interest to ecologists include the diversity, distribution, amount (biomass), and number (population) of particular organisms, as well as cooperation and competition between organisms, both within and among ecosystems. Ecosystems are composed of dynamically interacting parts including organisms, the communities they make up, and the non-living components of their environment. Ecosystem processes, such as primary production, pedogenesis, nutrient cycling, and various niche construction activities, regulate the flux of energy and matter through an environment. These processes are sustained by organisms with specific life history traits, and the variety of organisms is called biodiversity. Biodiversity, which refers to the varieties of species, genes, and ecosystems, enhances certain ecosystem services.

Ecology is not synonymous with environment, environmentalism, natural history, or environmental science. It is closely related to evolutionary biology, genetics, and ethology. An important focus for ecologists is to improve the understanding of how biodiversity affects ecological function. Ecologists seek to explain:

- Life processes, interactions, and adaptations
- The movement of materials and energy through living communities
- The successional development of ecosystems
- The abundance and distribution of organisms and biodiversity in the context of the environment.

There are many practical applications of ecology in conservation biology, wetland management, natural resource management (agroecology, agriculture, forestry, agroforestry, fisheries), city planning (urban ecology), community health, economics, basic and applied science, and human social interaction (human ecology). For example, the Circles of Sustainability approach treats ecology as more than the environment 'out there'. It is not treated as separate from humans. Organisms (including humans) and resources compose ecosystems which, in turn, maintain biophysicalfeedback mechanisms that moderate processes acting on living (biotic) and non-living (abiotic) components of the planet. Ecosystems sustain life-supporting functions and produce natural capital like biomass production (food, fuel, fiber, and medicine), the regulation of climate, global biogeochemical cycles, water filtration, soil formation, erosion control, flood protection, and many other natural features of scientific, historical, economic, or intrinsic value.

The word "ecology" ("Ökologie") was coined in 1866 by the German scientist Ernst Haeckel (1834–1919). Ecological thought is derivative of established currents in philosophy, particularly from ethics and politics. Ancient Greek philosophers such as Hippocrates and Aristotle laid the foundations of ecology in their studies on natural history. Modern ecology became a much more rigorous science in the late 19th century. Evolutionary concepts relating to adaptation and natural selection became the cornerstones of modern ecological theory.

DEFINITION OF ECOLOGY

The term oekologie (ecology) was coined in 1866 by the German biologist, Ernst Haeckel from the Greek oikos meaning "house" or "dwelling", and logos meaning "science" or "study". Thus, ecology is the "study of the household of nature". Haeckel intended it to encompass the study of an animal in relation to both the physical environment and other plants and animals with which it interacted.

A contemporary definition of ecology is:

The scientific study of the distribution and abundance of organisms and the interactions that determine distribution and abundance.

This definition encompasses not only the plants and animals that Haeckel recognized but microscopic organisms such as Bacteria, Archaea and protozoa, as well. The interactions that determine an organism's distribution and abundance are processes that include energy flow, growth, reproduction, predation, competition and many others.

Basic Ecology Definitions

To fully understand the science of ecology, there are some common terms that must be defined. The term environmentdescribes, in an unspecified way, the sum total of physical and biotic conditions that influence an organism

(Kendeigh, 1961). The subset of the planet earth environment into which life penetrates is termed the biosphere. With respect to the planet earth, the biosphere penetrates only a limited distance into the rock beneath the land and the oceans, and a limited distance out away from the planet towards space. All human effort so far has failed to demonstrate that the biosphere extends beyond these limits or that other biospheres exist elsewhere in the universe. We cannot therefore conclude that they do not exist, only that we know nothing of that existence. Ecosystem is perhaps the most widely used term in ecology. It is defined as the system of organisms and physical factors under study or consideration. Although the boundaries of ecosystems are sometimes quite difficult to define in nature, ecosystems—however bounded— comprise the basic units of that nature (T ansley, 1935). Habitat is generally considered by biologists to be the physical conditions that surround a species, or species population, or assemblage of species, or community (Clements and Shelford, 1939).

The basic physical units of the biosphere are the lithosphere (the land), hydrosphere (the water), and atmosphere (the air). Apparently there is no permanent biota of the atmosphere, although insects and birds among others utilize that environment extensively (Hesse, et al., 1951). These basic units are easily recognized in any landscape, as for example shown in the scene at right from Belarus in Europe.

HISTORICAL BACKGROUND

Ecology had no firm beginnings. It evolved from the natural history of the ancient Greeks, particularly Theophrastus, a friend and associate of Aristotle. Theophrastus first described the interrelationships between organisms and between organisms and their nonliving environment. Later foundations for modern ecology were laid in the early work of plant and animal physiologists.

In the early and mid-1900s two groups of botanists, one in Europe and the other in the United States, studied plant communities from two different points of view. The European botanists concerned themselves with the study of the composition, structure, and distribution of plant communities. The American botanists studied the development of plant communities, or succession (see community ecology: Ecological succession). Both plant and animal ecology developed separately until American biologists emphasized the interrelation of both plant and animal communities as a biotic whole.

During the same period, interest in population dynamics developed. The study of population dynamics received special impetus in the early 19th century, after the English economist Thomas Malthus called attention to the conflict between expanding populations and the capability of Earth to supply food. In the 1920s the American zoologist Raymond Pearl, the American chemist and statistician Alfred J. Lotka, and the Italian mathematician Vito Volterra developed mathematical foundations for the study of populations, and these studies led to experiments on the interaction of predators and prey, competitive relationships between species, and the regulation of populations. Investigations of the influence of behaviour on populations were stimulated by the recognition in 1920 of territoriality in nesting birds. Concepts of instinctive and aggressive behaviour were developed by the Austrian zoologist Konrad Lorenz and the Dutch-born British zoologist Nikolaas Tinbergen, and the role of social behaviour in the regulation of populations was explored by the British zoologist Vero Wynne-Edwards. (See population ecology.)

While some ecologists were studying the dynamics of communities and populations, others were concerned with energy budgets. In 1920 August Thienemann, a German freshwater biologist, introduced the concept of trophic, or feeding, levels (see trophic level), by which the energy of food is transferred through a series of organisms, from green plants (the producers) up to several levels of animals (the consumers). An English animal ecologist, Charles Elton (1927), further developed this approach with the concept of ecological niches and pyramids of numbers. In the 1930s, American freshwater biologists Edward Birge and Chancey Juday, in measuring the energy budgets of lakes, developed the idea of primary productivity, the rate at which food energy is generated,or fixed, by photosynthesis. In 1942 Raymond L. Lindeman of the United States developed the trophic-dynamic concept of ecology, which details the flow of energy through the ecosystem. Quantified field studies of energy flow through ecosystems were further developed by the brothers Eugene Odum and Howard Odum of the United States; similar early work on the cycling of nutrients was done by J.D. Ovington of England and Australia. (See community ecology: Trophic pyramids and the flow of energy; biosphere: The flow of energy and nutrient cycling.)

The study of both energy flow and nutrient cycling was stimulated by the development of new materials and techniques—radioisotope tracers, microcalorimetry , computer science, and applied mathematics—that enabled ecologists to label, track, and measure the movement of particular nutrients and energy through ecosystems. These modern methods (see below Methods in ecology) encouraged a new stage in the development of ecology—systems ecology , which is concerned with the structure and function of ecosystems.

Scope of Ecology

The scope of ecology contains a wide array of interacting levels of organization spanning micro-level (e.g., cells) to a planetary scale (e.g., biosphere) phenomena. Ecosystems, for example, contain abiotic resources and interacting life forms (i.e., individual organisms that aggregate into populations which aggregate into distinct ecological communities). Ecosystems are dynamic, they do not always follow a linear successional path, but they are always changing, sometimes rapidly and sometimes so slowly that it can take thousands of years for ecological processes to bring about certain successional stages of a forest. An ecosystem's area can vary greatly, from tiny to vast. A single tree is of little consequence to the classification of a forest ecosystem, but critically relevant to organisms living in and on it. Several generations of an aphid population can exist over the lifespan of a single leaf. Each of those aphids, in turn, support diverse bacterial communities. The nature of connections in ecological communities cannot be explained by knowing the details of each species in isolation, because the emergent pattern is neither revealed nor predicted until the ecosystem is studied as an integrated whole. Some ecological principles, however, do exhibit collective properties where the sum of the components explain the properties of the whole, such as birth rates of a population being equal to the sum of individual births over a designated time frame.

Hierarchy

System behaviors must first be arrayed into different levels of organization. Behaviors corresponding to higher levels occur at slow rates. Conversely, lower organizational levels exhibit rapid rates. For example, individual tree leaves respond rapidly to momentary changes in light intensity, CO concentration, and the like. The growth of the tree responds more slowly and integrates these short-term changes.

O'Neill et al. (1986)

The scale of ecological dynamics can operate like a closed system, such as aphids migrating on a single tree, while at the same time remain open with regard to broader scale influences, such as atmosphere or climate. Hence, ecologists classify ecosystems hierarchically by analyzing data collected from finer scale units, such as vegetation associations, climate, and soil types, and integrate this information to identify emergent patterns of uniform organization and processes that operate on local to regional, landscape, and chronological scales.

To structure the study of ecology into a conceptually manageable framework, the biological world is organized into a nested hierarchy, ranging in scale from genes, to cells, to tissues, to organs, to organisms, to species, to populations, to communities, to ecosystems, to biomes, and up to the level of the biosphere. This framework forms a panarchyand exhibits non-linear behaviors; this means that "effect and cause are disproportionate, so that small changes to critical variables, such as the number of nitrogen fixers, can lead to disproportionate, perhaps irreversible, changes in the system properties."

Biodiversity

Biodiversity (an abbreviation of "biological diversity") describes the diversity of life from genes to ecosystems and spans every level of biological organization. The term has several interpretations, and there are many ways to index, measure, characterize, and represent its complex organization. Biodiversity includes species diversity, ecosystem diversity, and genetic diversity and scientists are interested in the way that this diversity affects the complex ecological processes operating at and among these respective levels. Biodiversity plays an important role in ecosystem services which by definition maintain and improve human quality of life. Conservation priorities and management techniques require different approaches and considerations to address the full ecological scope of biodiversity. Natural capital that supports populations is critical for maintaining ecosystem services and species migration (e.g., riverine fish runs and avian insect control) has been implicated as one mechanism by which those service losses are experienced. An understanding of biodiversity has practical applications for species and ecosystem-level conservation planners as they make management recommendations to consulting firms, governments, and industry.

Habitat

The habitat of a species describes the environment over which a species is known to occur and the type of community that is formed as a result. More specifically, "habitats can be defined as regions in environmental space that are composed of multiple dimensions, each representing a biotic or abiotic environmental variable; that is, any component or characteristic of the environment related directly (e.g. forage biomass and quality) or indirectly (e.g. elevation) to the use of a location by the animal." For example, a habitat might be an aquatic or terrestrial environment that can be further categorized as a montane or alpine ecosystem. Habitat shifts provide important evidence of competition in nature where one population changes relative to the habitats that most other individuals of the species occupy. For example, one population of a species of tropical lizards (Tropidurus hispidus) has a flattened body relative to the main populations that live in open savanna. The population that lives in an isolated rock outcrop hides in crevasses where its flattened body offers a selective advantage. Habitat shifts also occur in the developmental life history of amphibians, and in insects that transition from aquatic to terrestrial habitats. Biotope and habitat are sometimes used interchangeably, but the former applies to a community's environment, whereas the latter applies to a species' environment.

Additionally, some species are ecosystem engineers, altering the environment within a localized region. For instance, beavers manage water levels by building dams which improves their habitat in a landscape.

Niche

Definitions of the niche date back to 1917, but G. Evelyn Hutchinson made conceptual advances in 1957 by introducing a widely adopted definition: "the set of biotic and abiotic conditions in which a species is able to persist and maintain stable population sizes." The ecological niche is a central concept in the ecology of organisms and is sub-divided into the fundamental and the realized niche. The fundamental niche is the set of environmental conditions under which a species is able to persist. The realized niche is the set of environmental plus ecological conditions under which a species persists. The Hutchinsonian niche is defined more technically as a "Euclidean hyperspace whose dimensions are defined as environmental variables and whose size is a function of the number of values that the environmental values may assume for which an organism has positive fitness."

Biogeographical patterns and range distributions are explained or predicted through knowledge of a species' traits and niche requirements. Species have functional traits that are uniquely adapted to the ecological niche. A trait is a measurable property, phenotype, or characteristic of an organism that may influence its survival. Genes play an important role in the interplay of development and environmental expression of traits. Resident species evolve traits that are fitted to the selection pressures of their local environment. This tends to afford them a competitive advantage and discourages similarly adapted species from having an overlapping geographic range. The competitive exclusion principle states that two species cannot coexist indefinitely by living off the same limiting resource; one will always out-compete the other. When similarly adapted species overlap geographically, closer inspection reveals subtle ecological differences in their habitat or dietary requirements. Some models and empirical studies, however, suggest that disturbances can stabilize the co-evolution and shared niche occupancy of similar species inhabiting species-rich communities. The habitat plus the niche is called the ecotope, which is defined as the full range of environmental and biological variables affecting an entire species.

Niche construction

Organisms are subject to environmental pressures, but they also modify their habitats. The regulatory feedback between organisms and their environment can affect conditions from local (e.g., a beaver pond) to global scales, over time and even after death, such as decaying logs or silica skeleton deposits from marine organisms. The process and concept of ecosystem engineering is related to niche construction, but the former relates only to the physical modifications of the habitat whereas the latter also considers the evolutionary implications of physical changes to the environment and the feedback this causes on the process of natural selection. Ecosystem engineers are defined as: "organisms that directly or indirectly modulate the availability of resources to other species, by causing physical state changes in biotic or abiotic materials. In so doing they modify, maintain and create habitats."

The ecosystem engineering concept has stimulated a new appreciation for the influence that organisms have on the ecosystem and evolutionary process. The term "niche construction" is more often used in reference to the under-appreciated feedback mechanisms of natural selection imparting forces on the abiotic niche. An example of natural selection through ecosystem engineering occurs in the nests of social insects, including ants, bees, wasps, and termites. There is an emergent homeostasis or homeorhesis in the structure of the nest that regulates, maintains and defends the physiology of the entire colony. Termite mounds, for example, maintain a constant internal temperature through the design of air-conditioning chimneys. The structure of the nests themselves are subject to the forces of natural selection. Moreover, a nest can survive over successive generations, so that progeny inherit both genetic material and a legacy niche that was constructed before their time.

Biome

Biomes are larger units of organization that categorize regions of the Earth's ecosystems, mainly according to the structure and composition of vegetation. There are different methods to define the continental boundaries of biomes dominated by different functional types of vegetative communities that are limited in distribution by climate, precipitation, weather and other environmental variables. Biomes include tropical rainforest, temperate broadleaf and mixed forest, temperate deciduous forest, taiga, tundra, hot desert, and polar desert. Other researchers have recently categorized other biomes, such as the human and oceanic microbiomes. To a microbe, the human body is a habitat and a landscape. Microbiomes were discovered largely through advances in molecular genetics, which have revealed a hidden richness of microbial diversity on the planet. The oceanic microbiome plays a significant role in the ecological biogeochemistry of the planet's oceans.

Biosphere

The largest scale of ecological organization is the biosphere: the total sum of ecosystems on the planet. Ecological relationships regulate the flux of energy, nutrients, and climate all the way up to the planetary scale. For example, the dynamic history of the planetary atmosphere's CO and O composition has been affected by the biogenic flux of gases coming from respiration and photosynthesis, with levels fluctuating over time in relation to the ecology and evolution of plants and animals.Ecological theory has also been used to explain self-emergent regulatory phenomena at the planetary scale: for example, the Gaia hypothesis is an example of holismapplied in ecological theory. The Gaia hypothesis states that there is an emergent feedback loop generated by the metabolism of living organisms that maintains the core temperature of the Earth and atmospheric conditions within a narrow self-regulating range of tolerance.

Individual ecology

Understanding traits of individual organisms helps explain patterns and processes at other levels of organization including populations, communities, and ecosystems. Several areas of ecology of evolution that focus on such traits are life history theory, ecophysiology, metabolic theory of ecology, and Ethology. Examples of such traits include features of an organisms life cycle such as age to maturity, life span, or metabolic costs of reproduction. Other traits may be related to structure, such as the spines of a cactus or dorsal spines of a bluegill sunfish, or behaviors such as courtship displays or pair bonding. Other traits include emergent properties that are the result at least in part of interactions with the surrounding environment such as growth rate, resource uptake rate, winter, and deciduous vs. drought deciduous trees and shrubs.

One set of characteristics relate to body size and temperature. The metabolic theory of ecology provides a predictive qualitative set of relationships between an organism's body size and temperature and metabolic processes. In general, smaller, warmer organisms have higher metabolic rates and this results in a variety of predictions regarding individual somatic growth rates, reproduction and population growth rates, population size, and resource uptake rates.

The traits of organisms are subject to change through acclimation, development, and evolution. For this reason, individuals form a shared focus for ecology and for evolutionary ecology.

METAPOPULATIONS AND MIGRATION

The concept of metapopulations was defined in 1969 as "a population of populations which go extinct locally and recolonize". Metapopulation ecology is another statistical approach that is often used in conservation research. Metapopulation models simplify the landscape into patches of varying levels of quality,and metapopulations are linked by the migratory behaviours of organisms. Animal migration is set apart from other kinds of movement; because, it involves the seasonal departure and return of individuals from a habitat. Migration is also a population-level phenomenon, as with the migration routes followed by plants as they occupied northern post-glacial environments. Plant ecologists use pollen records that accumulate and stratify in wetlands to reconstruct the timing of plant migration and dispersal relative to historic and contemporary climates. These migration routes involved an expansion of the range as plant populations expanded from one area to another. There is a larger taxonomy of movement, such as commuting, foraging, territorial behaviour, stasis, and ranging. Dispersal is usually distinguished from migration; because, it involves the one way permanent movement of individuals from their birth population into another population.

In metapopulation terminology, migrating individuals are classed as emigrants (when they leave a region) or immigrants (when they enter a region), and sites are classed either as sources or sinks. A site is a generic term that refers to places where ecologists sample populations, such as ponds or defined sampling areas in a forest. Source patches are productive sites that generate a seasonal supply of juveniles that migrate to other patch locations. Sink patches are unproductive sites that only receive migrants; the population at the site will disappear unless rescued by an adjacent source patch or environmental conditions become more favourable. Metapopulation models examine patch dynamics over time to answer potential questions about spatial and demographic ecology. The ecology of metapopulations is a dynamic process of extinction and colonization. Small patches of lower quality (i.e., sinks) are maintained or rescued by a seasonal influx of new immigrants. A dynamic metapopulation structure evolves from year to year, where some patches are sinks in dry years and are sources when conditions are more favourable. Ecologists use a mixture of computer models and field studies to explain metapopulation structure.

Community ecology

Interspecific interactions such as predation are a key aspect of community ecology.

Community ecology is the study of the interactions among a collections of species that inhabit the same geographic area. Community ecologists study the determinants of patterns and processes for two or more interacting species. Research in community ecology might measure species diversity in grasslands in relation to soil fertility. It might also include the analysis of predator-prey dynamics, competition among similar plant species, or mutualistic interactions between crabs and corals.

Ecosystem ecology

These ecosystems, as we may call them, are of the most various kinds and sizes. They form one category of the multitudinous physical systems of the universe, which range from the universe as a whole down to the atom.

Tansley (1935)

A riparian forest in the White Mountains, New Hampshire (USA) is an example of ecosystem ecology

Ecosystems may be habitats within biomes that form an integrated whole and a dynamically responsive system having both physical and biological complexes. Ecosystem ecology is the science of determining the fluxes of materials (e.g. carbon, phosphorus) between different pools (e.g., tree biomass, soil organic material). Ecosystem ecologist attempt to determine the underlying causes of these fluxes. Research in ecosystem ecology might measure primary production (g C/m^2) in a wetland in relation to decomposition and consumption rates (g $C/m^2/y$). This requires an understanding of the community connections between plants (i.e., primary producers) and the decomposers (e.g., fungi and bacteria), The underlying concept of ecosystem can be traced back to 1864 in the published work of George Perkins Marsh("Man and Nature"). Within an ecosystem, organisms are linked to the physical and biological components of their environment to which they are adapted. Ecosystems are complex adaptive systems where the interaction of life processes form self-organizing patterns across different scales of time and space. Ecosystems are broadly categorized as terrestrial, freshwater, atmospheric, or marine. Differences stem from the nature of the unique physical environments that shapes the biodiversity within each. A more recent addition to ecosystem ecology are technoecosystems, which are affected by or primarily the result of human activity.

Food webs

A food web is the archetypal ecological network. Plants capture solar energy and use it to synthesize simple sugars during photosynthesis. As plants grow, they accumulate nutrients and are eaten by grazing herbivores, and the energy is transferred through a chain of organisms by consumption. The simplified linear feeding pathways that move from a basal trophic species to a top consumer is called the food chain. The larger interlocking pattern of food chains in an ecological community creates a complex food web. Food webs are a type of concept map or a heuristic device that is used to illustrate and study pathways of energy and material flows.

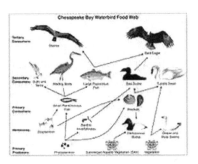

Generalized food web of waterbirds from Chesapeake Bay

Food webs are often limited relative to the real world. Complete empirical measurements are generally restricted to a specific habitat, such as a cave or a pond, and principles gleaned from food web microcosm studies are extrapolated to larger systems. Feeding relations require extensive investigations into the gut contents of organisms, which can be difficult to decipher, or stable isotopes can be used to trace the flow of nutrient diets and energy through a food web. Despite these limitations, food webs remain a valuable tool in understanding community ecosystems.

Food webs exhibit principles of ecological emergence through the nature of trophic relationships: some species have many weak feeding links (e.g., omnivores) while some are more specialized with fewer stronger feeding links (e.g., primary predators). Theoretical and empirical studies identify non-random emergent patterns of few strong and many weak linkages that explain how ecological communities remain stable over time. Food webs are composed of subgroups where members in a community are linked by strong interactions, and the weak interactions occur between these subgroups. This increases food web stability. Step by step lines or relations are drawn until a web of life is illustrated.

Trophic levels

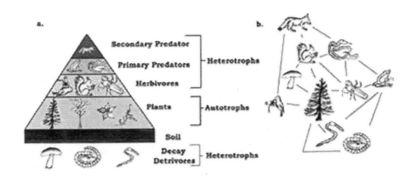

A trophic pyramid (a) and a food-web (b) illustrating ecological relationshipsamong creatures that are typical of a northern boreal terrestrial ecosystem. The trophic pyramid roughly represents the biomass (usually measured as total dry-weight) at each level. Plants generally have the greatest biomass. Names of trophic categories are shown to the right of the pyramid. Some ecosystems, such as many wetlands, do not organize as a strict pyramid, because aquatic plants are not as productive as long-lived terrestrial plants such as trees. Ecological trophic pyramids are typically one of three kinds: 1) pyramid of numbers, 2) pyramid of biomass, or 3) pyramid of energy.

A trophic level (from Greek troph, ôñïöÞ, trophç, meaning "food" or "feeding") is "a group of organisms acquiring a considerable majority of its energy from the adjacent level nearer the abiotic source." Links in food webs primarily connect feeding relations or trophism among species. Biodiversity within ecosystems can be organized into trophic pyramids, in which the vertical dimension represents feeding relations that become further removed from the base of the food chain up toward top predators, and the horizontal dimension represents the abundance or biomass at each level. When the relative abundance or biomass of each species is sorted into its respective trophic level, they naturally sort into a 'pyramid of numbers'.

Species are broadly categorized as autotrophs (or primary producers), heterotrophs (or consumers), and Detritivores (or decomposers). Autotrophs are organisms that produce their own food (production is greater than respiration) by photosynthesis or chemosynthesis. Heterotrophs are organisms that must feed on others for nourishment and energy (respiration exceeds production). Heterotrophs can be further sub-divided into different functional groups, including primary consumers (strict herbivores), secondary consumers (carnivorous predators that feed exclusively on herbivores), and tertiary consumers (predators that feed on a mix of herbivores and predators). Omnivores do not fit neatly into a functional category because they eat both plant and animal tissues. It has been suggested that omnivores have a greater functional influence as predators, because compared to herbivores, they are relatively inefficient at grazing.

Trophic levels are part of the holistic or complex systems view of ecosystems. Each trophic level contains unrelated species that are grouped together because they share common ecological functions, giving a macroscopic view of the system. While the notion of trophic levels provides insight into energy flow and top-down control within food webs, it is troubled by the prevalence of omnivory in real ecosystems. This has led some ecologists to "reiterate that the notion that species clearly aggregate into discrete, homogeneous trophic levels is fiction." Nonetheless, recent studies have shown that real trophic levels do exist, but "above the herbivore trophic level, food webs are better characterized as a tangled web of omnivores."

Keystone species

A keystone species is a species that is connected to a disproportionately large number of other species in the food-web. Keystone species have lower levels of biomass in the trophic pyramid relative to the importance of their role. The many connections that a keystone species holds means that it maintains the organization and structure of entire communities. The loss of a keystone species results in a range of dramatic cascading effects that alters trophic dynamics, other food web connections, and can cause the extinction of other species.

Sea otters (Enhydra lutris) are commonly cited as an example of a keystone species; because, they limit the density of sea urchins that feed on kelp. If sea otters are removed from the system, the urchins graze until the kelp beds disappear, and this has a dramatic effect on community structure. Hunting of sea otters, for example, is thought to have led indirectly to the extinction of the Steller's sea cow (Hydrodamalis gigas). While the keystone species concept has been used extensively as a conservation tool, it has been criticized for being poorly defined from an operational stance. It is difficult to experimentally determine what species may hold a keystone role in each ecosystem. Furthermore, food web theory suggests that keystone species may not be common, so it is unclear how generally the keystone species model can be applied.

Methods In Ecology

Because ecologists work with living systems possessing numerous variables, the scientific techniques used by physicists, chemists, mathematicians, and engineers require modification for use in ecology. Moreover, the techniques are not as easily applied in ecology, nor are the results as precise as those obtained in other sciences. It is relatively simple, for example, for a physicist to measure gain and loss of heat from metals or other inanimate objects, which possess certain constants of conductivity, expansion, surface features, and the like. To determine the heat exchange between an animal and its environment, however, a physiological ecologist is confronted with an array of almost unquantifiable variables and with the formidable task of gathering the numerous data and analyzing them. Ecological measurements may never be as precise or subject to the same ease of analysis as measurements in physics, chemistry, or certain quantifiable areas of biology.

In spite of these problems, various aspects of the environment can be determined by physical and chemical means, ranging from simple chemical identifications and physical measurements to the use of sophisticated mechanical apparatus. The development of biostatistics (statistics applied to the analysis of biological data), the elaboration of proper experimental design, and improved sampling methods now permit a quantified statistical approach to the study of ecology. Because of the extreme difficulties of controlling environmental variables in the field, studies involving the use of experimental design are largely confined to the laboratory and to controlled field experiments designed to test the effects of only one variable or several variables. The use of statistical procedures and computer models based on data obtained from the field provide insights into population interactions and ecosystem functions.

Mathematical programming models are becoming increasingly important in applied ecology, especially in the management of natural resources and agricultural problems having an ecological basis.

Controlled environmental chambers enable experimenters to maintain plants and animals under known conditions of light, temperature, humidity, and day length so that the effects of each variable (or combination of variables) on the organism can be studied. Biotelemetry and other electronic tracking equipment, which allow the movements and behaviour of free-ranging organisms to be followed remotely, can provide rapid sampling of populations. Radioisotopes are used for tracing the pathways of nutrients through ecosystems, for determining the time and extent of transfer of energy and nutrients through the different components of the ecosystem, and for the determination of food chains. The use of laboratory microcosms—aquatic and soil micro-ecosystems, consisting of biotic and nonbiotic material from natural ecosystems, held under conditions similar to those found in the field—are useful in determining rates of nutrient cycling, ecosystem development, and other functional aspects of ecosystems. Microcosms enable the ecologist to duplicate experiments and to perform experimental manipulation on them.

ECOLOGICAL COMPLEXITY

Complexity is understood as a large computational effort needed to piece together numerous interacting parts exceeding the iterative memory capacity of the human mind. Global patterns of biological diversity are complex. This biocomplexity stems from the interplay among ecological processes that operate and influence patterns at different scales that grade into each other, such as transitional areas or ecotones spanning landscapes. Complexity stems from the interplay among levels of biological organization as energy, and matter is integrated into larger units that superimpose onto the smaller parts. "What were wholes on one level become parts on a higher one." Small scale patterns do not necessarily explain large scale phenomena, otherwise captured in the expression (coined by Aristotle) 'the sum is greater than the parts'.

"Complexity in ecology is of at least six distinct types: spatial, temporal, structural, process, behavioral, and geometric." From these principles, ecologists have identified emergent and self-organizing phenomena that operate at different environmental scales of influence, ranging from molecular to planetary, and these require different explanations at each integrative level. Ecological complexity relates to the dynamic resilience of ecosystems that transition to multiple shifting steady-states directed by random fluctuations of history. Long-term ecological studies provide important track records to better understand the complexity and resilience of ecosystems over longer temporal and broader spatial scales. These studies are managed by the International Long Term Ecological Network (LTER). The longest experiment in existence is the Park Grass Experiment, which was initiated in 1856. Another example is the Hubbard Brook study, which has been in operation since 1960.

Holism

Holism remains a critical part of the theoretical foundation in contemporary ecological studies. Holism addresses the biological organization of life that self-organizesinto layers of emergent whole systems that function according to non-reducible properties. This means that higher order patterns of a whole functional system, such as an ecosystem, cannot be predicted or understood by a simple summation of the parts. "New properties emerge because the components interact, not because the basic nature of the components is changed."

Ecological studies are necessarily holistic as opposed to reductionistic. Holism has three scientific meanings or uses that identify with ecology: 1) the mechanistic complexity of ecosystems, 2) the practical description of patterns in quantitative reductionist terms where correlations may be identified but nothing is understood about the causal relations without reference to the whole system, which leads to 3) a metaphysical hierarchy whereby the causal relations of larger systems are understood without reference to the smaller parts. Scientific holism differs from mysticism that has appropriated the same term. An example of metaphysical holism is identified in the trend of increased exterior thickness in shells of different species. The reason for a thickness increase can be understood through reference to principles of natural selection via predation without need to reference or understand the biomolecular properties of the exterior shells.

Relation to evolution

Ecology and evolutionary biology are considered sister disciplines of the life sciences. Natural selection, life history, development, adaptation, populations, and inheritance are examples of concepts that thread equally into ecological and evolutionary theory. Morphological, behavioural, and genetic traits, for example, can be mapped onto evolutionary trees to study the historical development of a species in relation to their functions and roles in different ecological circumstances. In this framework, the analytical tools of ecologists and evolutionists overlap as they organize, classify, and investigate life through common systematic principals, such as phylogenetics or the Linnaean system of taxonomy. The two disciplines often appear together, such as in the title of the journal Trends in Ecology and Evolution. There is no sharp boundary separating ecology from evolution, and they differ more in their areas of applied focus. Both disciplines discover and explain emergent and unique properties and processes operating across different spatial or temporal scales of organization. While the boundary between ecology and evolution is not always clear, ecologists study the abiotic and biotic factors that influence evolutionary processes, and evolution can be rapid, occurring on ecological timescales as short as one generation.

Behavioural ecology

All organisms can exhibit behaviours. Even plants express complex behaviour, including memory and communication. Behavioural ecology is the study of an organism's behaviour in its environment and its ecological and evolutionary implications. Ethology is the study of observable movement or behaviour in animals. This could include investigations of motile sperm of plants, mobile phytoplankton, zooplanktonswimming toward the female egg, the cultivation of fungi by weevils, the mating dance of a salamander, or social gatherings of amoeba.

Adaptation is the central unifying concept in behavioural ecology. Behaviours can be recorded as traits and inherited in much the same way that eye and hair colour can. Behaviours can evolve by means of natural selection as adaptive traits conferring functional utilities that increases reproductive fitness.

Predator-prey interactions are an introductory concept into food-web studies as well as behavioural ecology. Prey species can exhibit different kinds of behavioural adaptations to predators, such as avoid, flee, or defend. Many prey species are faced with multiple predators that differ in the degree of danger posed. To be adapted to their environment and face predatory threats, organisms must balance their energy budgets as they invest in different aspects of their life history, such as growth, feeding, mating, socializing, or modifying their habitat. Hypotheses posited in behavioural ecology are generally based on adaptive principles of conservation, optimization, or efficiency. For example, "[t]he threat-sensitive predator avoidance hypothesis predicts that prey should assess the degree of threat posed by different predators and match their behaviour according to current levels of risk" or "[t]he optimal flight initiation distance occurs where expected postencounter fitness is maximized, which depends on the prey's initial fitness, benefits obtainable by not fleeing, energetic escape costs, and expected fitness loss due to predation risk."

Symbiosis: Leafhoppers (Eurymela fenestrata) are protected by ants(Iridomyrmex purpureus) in a symbioticrelationship. The ants protect the leafhoppers from predators and in return the leafhoppers feeding on plants exude honeydew from their anus that provides energy and nutrients to tending ants.

Elaborate sexual displays and posturing are encountered in the behavioural ecology of animals. The birds-of-paradise, for example, sing and display elaborate ornaments during courtship. These displays serve a dual purpose of signalling healthy or well-adapted individuals and desirable genes. The displays are driven by sexual selection as an advertisement of quality of traits among suitors.

Cognitive ecology

Cognitive ecology integrates theory and observations from evolutionary ecology and neurobiology, primarily cognitive science, in order to understand the effect that animal interaction with their habitat has on their cognitive systems and how those systems restrict behavior within an ecological and evolutionary framework. "Until recently, however, cognitive scientists have not paid sufficient attention to the fundamental fact that cognitive traits evolved under particular natural settings. With consideration of the selection pressure on cognition, cognitive ecology can contribute intellectual coherence to the multidisciplinary study of cognition." As a study involving the 'coupling' or interactions between organism and environment, cognitive ecology is closely related to enactivism, a field based upon the view that "...we must see the organism and environment as bound together in reciprocal specification and selection...".

Social ecology

Social ecological behaviours are notable in the social insects, slime moulds, social spiders, human society, and naked mole-rats where eusocialism has evolved. Social behaviours include reciprocally beneficial behaviours among kin and nest mates and evolve from kin and group selection. Kin selection explains altruism through genetic relationships, whereby an altruistic behaviour leading to death is rewarded by the survival of genetic copies distributed among surviving relatives. The social insects, including ants, bees, and wasps are most famously studied for this type of relationship because the male drones are clones that share the same genetic make-up as every other male in the colony. In contrast, group selectionists find examples of altruism among non-genetic relatives and explain this through selection acting on the group; whereby, it becomes selectively advantageous for groups if their members express altruistic behaviours to one another. Groups with predominantly altruistic members beat groups with predominantly selfish members.

Coevolution

Ecological interactions can be classified broadly into a host and an associate relationship. A host is any entity that harbours another that is called the associate. Relationships within a species that are mutually or reciprocally beneficial are called mutualisms. Examples of mutualism include fungus-growing ants employing agricultural symbiosis, bacteria living in the guts of insects and other organisms, the fig wasp and yucca moth pollination complex, lichens with fungi and photosynthetic algae, and corals with photosynthetic algae. If there is a physical connection between host and associate, the relationship is called symbiosis. Approximately 60% of all plants, for example, have a symbiotic relationship with arbuscular mycorrhizal fungi living in their roots forming an exchange network of carbohydrates for mineral nutrients.

Indirect mutualisms occur where the organisms live apart. For example, trees living in the equatorial regions of the planet supply oxygen into the atmosphere that sustains species living in distant polar regions of the planet. This relationship is called commensalism; because, many others receive the benefits of clean air at no cost or harm to trees supplying the oxygen. If the associate benefits while the host suffers, the relationship is called parasitism. Although parasites impose a cost to their host (e.g., via damage to their reproductive organs or propagules, denying the services of a beneficial partner), their net effect on host fitness is not necessarily negative and, thus, becomes difficult to forecast. Co-evolution is also driven by competition among species or among members of the same species under the banner of reciprocal antagonism, such as grasses competing for growth space. The Red Queen Hypothesis, for example, posits that parasites track down and specialize on the locally common genetic defense systems of its host that drives the evolution of sexual reproduction to diversify the genetic constituency of populations responding to the antagonistic pressure.

Parasitism: A harvestman arachnid being parasitized by mites. The harvestman is being consumed, while the mites benefit from traveling on and feeding off of their host.

Biogeography

Biogeography (an amalgamation of biology and geography) is the comparative study of the geographic distribution of organisms and the corresponding evolution of their traits in space and time. The Journal of Biogeographywas established in 1974. Biogeography and ecology share many of their disciplinary roots. For example, the theory of island biogeography, published by the Robert MacArthur and Edward O. Wilson in 1967 is considered one of the fundamentals of ecological theory.

Biogeography has a long history in the natural sciences concerning the spatial distribution of plants and animals. Ecology and evolution provide the explanatory context for biogeographical studies. Biogeographical patterns result from ecological processes that influence range distributions, such as migration and dispersal. and from historical processes that split populations or species into different areas. The biogeographic processes that result in the natural splitting of species explains much of the modern distribution of the Earth's biota. The splitting of lineages in a species is called vicariance biogeography and it is a sub-discipline of biogeography. There are also practical applications in the field of biogeography concerning ecological systems and processes. For example, the range and distribution of biodiversity and invasive species responding to climate change is a serious concern and active area of research in the context of global warming.

r/K selection theory

A population ecology concept is r/K selection theory, one of the first predictive models in ecology used to explain life-history evolution. The premise behind the r/K selection model is that natural selection pressures change according to population density. For example, when an island is first colonized, density of individuals is low. The initial increase in population size is not limited by competition, leaving an abundance of available resources for rapid population growth. These early phases of population growth experience density-independent forces of natural selection, which is called r-selection. As the population becomes more crowded, it approaches the island's carrying capacity, thus forcing individuals to compete more heavily for fewer available resources. Under crowded conditions, the population experiences density-dependent forces of natural selection, called K-selection.

In the r/K-selection model, the first variable r is the intrinsic rate of natural increase in population size and the second variable K is the carrying capacity of a population. Different species evolve different life-history strategies spanning a continuum between these two selective forces. An r-selected species is one that has high birth rates, low levels of parental investment, and high rates of mortality before individuals reach maturity. Evolution favours high rates of fecundity in r-selected species. Many kinds of insects and invasive species exhibit r-selected characteristics. In contrast, a K-selected species has low rates of fecundity, high levels of parental investment in the young, and low rates of mortality as individuals mature. Humans and elephants are examples of species exhibiting K-selected characteristics, including longevity and efficiency in the conversion of more resources into fewer offspring.

Molecular ecology

The important relationship between ecology and genetic inheritance predates modern techniques for molecular analysis. Molecular ecological research became more feasible with the development of rapid and accessible genetic technologies, such as the polymerase chain reaction (PCR). The rise of molecular technologies and influx of research questions into this new ecological field resulted in the publication Molecular Ecology in 1992. Molecular ecology uses various analytical techniques to study genes in an evolutionary and ecological context. In 1994, John Avise also played a leading role in this area of science with the publication of his book, Molecular Markers, Natural

History and Evolution. Newer technologies opened a wave of genetic analysis into organisms once difficult to study from an ecological or evolutionary standpoint, such as bacteria, fungi, and nematodes. Molecular ecology engendered a new research paradigm for investigating ecological questions considered otherwise intractable. Molecular investigations revealed previously obscured details in the tiny intricacies of nature and improved resolution into probing questions about behavioural and biogeographical ecology. For example, molecular ecology revealed promiscuous sexual behaviour and multiple male partners in tree swallows previously thought to be socially monogamous. In a biogeographical context, the marriage between genetics, ecology, and evolution resulted in a new sub-discipline called phylogeography.

Human ecology

Ecology is as much a biological science as it is a human science. Human ecology is an interdisciplinary investigation into the ecology of our species. "Human ecology may be defined: (1) from a bioecological standpoint as the study of man as the ecological dominant in plant and animal communities and systems; (2) from a bioecological standpoint as simply another animal affecting and being affected by his physical environment; and (3) as a human being, somehow different from animal life in general, interacting with physical and modified environments in a distinctive and creative way. A truly interdisciplinary human ecology will most likely address itself to all three." The term was formally introduced in 1921, but many sociologists, geographers, psychologists, and other disciplines were interested in human relations to natural systems centuries prior, especially in the late 19th century.

The ecological complexities human beings are facing through the technological transformation of the planetary biome has brought on the Anthropocene. The unique set of circumstances has generated the need for a new unifying science called coupled human and natural systems that builds upon, but moves beyond the field of human ecology.Ecosystems tie into human societies through the critical and all encompassing life-supporting functions they sustain. In recognition of these functions and the incapability of traditional economic valuation methods to see the value in ecosystems, there has been a surge of interest in social-natural capital, which provides the means to put a value on the stock and use of information and materials stemming from ecosystem goods and services. Ecosystems produce, regulate, maintain, and supply services of critical necessity and beneficial to human health (cognitive and physiological), economies, and they even provide an information or reference function as a living library giving opportunities for science and cognitive development in children engaged in the complexity of the natural world. Ecosystems relate importantly to human ecology as they are the ultimate base foundation of global economics as every commodity, and the capacity for exchange ultimately stems from the ecosystems on Earth.

RESTORATION AND MANAGEMENT

Ecology is an employed science of restoration, repairing disturbed sites through human intervention, in natural resource management, and in environmental impact assessments. Edward O. Wilson predicted in 1992 that the 21st century "will be the era of restoration in ecology". Ecological science has boomed in the industrial investment of restoring ecosystems and their processes in abandoned sites after disturbance. Natural resource managers, in forestry, for example, employ ecologists to develop, adapt, and implement ecosystem based methods into the planning, operation, and restoration phases of land-use. Ecological science is used in the methods of sustainable harvesting, disease, and fire outbreak management, in fisheries stock management, for integrating land-use with protected areas and communities, and conservation in complex geo-political landscapes.

Relation to the environment

The environment of ecosystems includes both physical parameters and biotic attributes. It is dynamically interlinked, and contains resources for organisms at any time throughout their life cycle. Like ecology, the term environment has different conceptual meanings and overlaps with the concept of nature. Environment "includes the physical world, the social world of human relations and the built world of human creation." The physical environment is external to the level of biological organization under investigation, including abiotic factors such as temperature, radiation, light, chemistry, climate and geology. The biotic environment includes genes, cells, organisms, members of the same species (conspecifics) and other species that share a habitat.

The distinction between external and internal environments, however, is an abstraction parsing life and environment into units or facts that are inseparable in reality. There is an interpenetration of cause and effect between the environment and life. The laws of thermodynamics, for example, apply to ecology by means of its physical state. With an understanding of metabolic and thermodynamic principles, a complete accounting of energy and material flow can be traced through an ecosystem. In this way, the environmental and ecological relations are studied through reference to conceptually manageable and isolated material parts. After the effective environmental components are understood through reference to their causes; however, they conceptually link back together as an integrated whole, or holocoenoticsystem as it was once called. This is known as the dialectical approach to ecology. The dialectical approach examines the parts, but integrates the organism and the environment into a dynamic whole (or umwelt). Change in one ecological or environmental factor can concurrently affect the dynamic state of an entire ecosystem.

Disturbance and resilience

Ecosystems are regularly confronted with natural environmental variations and disturbances over time and geographic space. A disturbance is any process that removes biomass from a community, such as a fire, flood, drought, or predation. Disturbances occur over vastly different ranges in terms of magnitudes as well as distances and time periods, and are both the cause and product of natural fluctuations in death rates, species assemblages, and biomass densities within an ecological community. These disturbances create places of renewal where new directions emerge from the patchwork of natural experimentation and opportunity. Ecological resilience is a cornerstone theory in ecosystem management. Biodiversity fuels the resilience of ecosystems acting as a kind of regenerative insurance.

Metabolism and the early atmosphere

Metabolism – the rate at which energy and material resources are taken up from the environment, transformed within an organism, and allocated to maintenance, growth and reproduction – is a fundamental physiological trait.

Ernest et al.

The Earth was formed approximately 4.5 billion years ago. As it cooled and a crust and oceans formed, its atmosphere transformed from being dominated by hydrogen to one composed mostly of methane and ammonia. Over the next billion years, the metabolic activity of life transformed the atmosphere into a mixture of carbon dioxide, nitrogen, and water vapor. These gases changed the way that light from the sun hit the Earth's surface and greenhouse effects trapped heat. There were untapped sources of free energy within the mixture of reducing and oxidizing gasses that set the stage for primitive ecosystems to evolve and, in turn, the atmosphere also evolved.

The leaf is the primary site of photosynthesis in most plants.

Throughout history, the Earth's atmosphere and biogeochemical cycles have been in a dynamic equilibrium with planetary ecosystems. The history is characterized by periods of significant transformation followed by millions of years of stability. The evolution of the earliest organisms, likely anaerobic methanogen microbes, started the process by converting atmospheric hydrogen into methane (4H+ CO '! CH + 2HO). Anoxygenic photosynthesis reduced hydrogen concentrations and increased atmospheric methane, by converting hydrogen sulfide into water or other sulfur compounds (for example, 2HS + CO + hv '! CHO + HO + 2S). Early forms of fermentation also increased levels of atmospheric methane. The transition to an oxygen-dominant atmosphere (the Great Oxidation) did not begin until approximately 2.4–2.3 billion years ago, but photosynthetic processes started 0.3 to 1 billion years prior.

Radiation: heat, temperature and light

The biology of life operates within a certain range of temperatures. Heat is a form of energy that regulates temperature. Heat affects growth rates, activity, behaviour, and primary production. Temperature is largely dependent on the incidence of solar radiation. The latitudinal and longitudinal spatial variation of temperature greatly affects climates and consequently the distribution of biodiversity and levels of primary production in different ecosystems or biomes across the planet. Heat and temperature relate importantly to metabolic activity. Poikilotherms, for example, have a body temperature that is largely regulated and dependent on the temperature of the external environment. In contrast, homeotherms regulate their internal body temperature by expending metabolic energy.

There is a relationship between light, primary production, and ecological energy budgets. Sunlight is the primary input of energy into the planet's ecosystems. Light is composed of electromagnetic energy of different wavelengths. Radiant energy from the sun generates heat, provides photons of light measured as active energy in the chemical reactions of life, and also acts as a catalyst for genetic mutation. Plants, algae, and some bacteria absorb light and assimilate the energy through photosynthesis. Organisms capable of assimilating energy by photosynthesis or through inorganic fixation of HS are autotrophs. Autotrophs — responsible for primary production — assimilate light energy which becomes metabolically stored as potential energy in the form of biochemical enthalpic bonds.

PHYSICAL ENVIRONMENTS

Water

Diffusion of carbon dioxide and oxygen is approximately 10,000 times slower in water than in air. When soils are flooded, they quickly lose oxygen, becoming hypoxic (an environment with O concentration below 2 mg/liter) and eventually completely anoxic where anaerobic bacteria thrive among the roots. Water also influences the intensity and spectral composition of light as it reflects off the water surface and submerged particles. Aquatic plants exhibit a wide variety of morphological and physiological adaptations that allow them to survive, compete, and diversify in these environments. For example, their roots and stems contain large air spaces (aerenchyma) that regulate the efficient transportation of gases (for example, CO and O) used in respiration and photosynthesis. Salt water plants (halophytes) have additional specialized adaptations, such as the development of special organs for shedding salt and osmoregulating their internal salt (NaCl) concentrations, to live in estuarine, brackish, or oceanic environments. Anaerobic soil microorganisms in aquatic environments use nitrate, manganese ions, ferric ions, sulfate, carbon dioxide, and some organic compounds; other microorganisms are facultative anaerobes and use oxygen during respiration when the soil becomes drier. The activity of soil microorganisms and the chemistry of the water reduces the oxidation-reduction potentials of the water. Carbon dioxide, for example, is reduced to methane (CH) by methanogenic bacteria. The physiology of fish is also specially adapted to compensate for environmental salt levels through osmoregulation. Their gills form electrochemical gradients that mediate salt excretion in salt water and uptake in fresh water.

Gravity

The shape and energy of the land is significantly affected by gravitational forces. On a large scale, the distribution of gravitational forces on the earth is uneven and influences the shape and movement of tectonic plates as well as influencing geomorphic processes such as orogeny and erosion. These forces govern many of the geophysical properties and distributions of ecological biomes across the Earth. On the organismal scale, gravitational forces provide directional cues for plant and fungal growth (gravitropism), orientation cues for animal migrations, and influence the biomechanics and size of animals. Ecological traits, such as allocation of biomass in trees during growth are subject to mechanical failure as gravitational forces influence the position and structure of branches and leaves. The cardiovascular systems of animals are functionally adapted to overcome pressure and gravitational forces that change according to the features of organisms (e.g., height, size, shape), their behaviour (e.g., diving, running, flying), and the habitat occupied (e.g., water, hot deserts, cold tundra).

Pressure

Climatic and osmotic pressure places physiological constraints on organisms, especially those that fly and respire at high altitudes, or dive to deep ocean depths.These constraints influence vertical limits of ecosystems in the biosphere, as organisms are physiologically sensitive and adapted to atmospheric and osmotic water pressure differences. For example, oxygen levels decrease with decreasing pressure and are a limiting factor for life at higher altitudes. Water transportation by plants is another important ecophysiological process affected by osmotic pressure gradients. Water pressure in the depths of oceans requires that organisms adapt to these conditions. For example, diving animals such as whales, dolphins, and seals are specially adapted to deal with changes in sound due to water pressure differences. Differences between hagfish species provide another example of adaptation to deep-sea pressure through specialized protein adaptations.

Wind and turbulence

Turbulent forces in air and water affect the environment and ecosystem distribution, form and dynamics. On a planetary scale, ecosystems are affected by circulation patterns in the global trade winds. Wind power and the turbulent forces it creates can influence heat, nutrient, and biochemical profiles of ecosystems. For example, wind running over the surface of a lake creates turbulence, mixing the water column and influencing the environmental profile to create thermally layered zones, affecting how fish, algae, and other parts of the aquatic ecosystem are structured. Wind speed and turbulence also influence evapotranspiration rates and energy budgets in plants and animals. Wind speed, temperature and moisture content can vary as winds travel across different land features and elevations. For example, the westerlies come into contact with the coastal and interior mountains of western North America to produce a rain shadow on the leeward side of the mountain. The air expands and moisture condenses as the winds increase in elevation; this is called orographic lift and can cause precipitation. This environmental process produces spatial divisions in biodiversity, as species adapted to wetter conditions are range-restricted to the coastal mountain valleys and unable to migrate across the xeric ecosystems (e.g., of the Columbia Basin in western North America) to intermix with sister lineages that are segregated to the interior mountain systems.

Fire

Plants convert carbon dioxide into biomass and emit oxygen into the atmosphere. By approximately 350 million years ago (the end of the Devonian period), photosynthesis had brought the concentration of atmospheric oxygen above 17%, which allowed combustion to occur. Fire releases CO and converts fuel into ash and tar. Fire is a significant ecological parameter that raises many issues pertaining to its control and suppression. While the issue of fire in relation to ecology and plants has been recognized for a long time, Charles Cooper brought attention to the issue of forest fires in relation to the ecology of forest fire suppression and management in the 1960s.

Native North Americans were among the first to influence fire regimes by controlling their spread near their homes or by lighting fires to stimulate the production of herbaceous foods and basketry materials. Fire creates a heterogeneous ecosystem age and canopy structure, and the altered soil nutrient supply and cleared canopy structure opens new ecological niches for seedling establishment. Most ecosystems are adapted to natural fire cycles. Plants, for example, are equipped with a variety of adaptations to deal with forest fires. Some species (e.g., Pinus halepensis) cannot germinateuntil after their seeds have lived through a fire or been exposed to certain compounds from smoke. Environmentally triggered germination of seeds is called serotiny. Fire plays a major role in the persistence and resilience of ecosystems.

Soils

Soil is the living top layer of mineral and organic dirt that covers the surface of the planet. It is the chief organizing centre of most ecosystem functions, and it is of critical importance in agricultural science and ecology. The decomposition of dead organic matter (for example, leaves on the forest floor), results in soils containing minerals and nutrients that feed into plant production. The whole of the planet's soil ecosystems is called the pedosphere where a large biomass of the Earth's biodiversity organizes into trophic levels. Invertebrates that feed and shred larger leaves, for example, create smaller bits for smaller organisms in the feeding chain. Collectively, these organisms are the detritivores that regulate soil formation. Tree roots, fungi, bacteria, worms, ants, beetles, centipedes, spiders, mammals, birds, reptiles, amphibians, and other less familiar creatures all work to create the trophic web of life in soil ecosystems. Soils form composite phenotypes where inorganic matter is enveloped into the physiology of a whole community. As organisms feed and migrate through soils they physically displace materials, an ecological process called bioturbation. This aerates soils and stimulates heterotrophic growth and production. Soil microorganisms are influenced by and feed back into the trophic dynamics of the ecosystem. No single axis of causality can be discerned to segregate the biological from geomorphological systems in soils. Paleoecologicalstudies of soils places the origin for bioturbation to a time before the Cambrian period. Other events, such as the evolution of trees and the colonization of land in the Devonian period played a significant role in the early development of ecological trophism in soils.

Biogeochemistry and climate

Ecologists study and measure nutrient budgets to understand how these materials are regulated, flow, and recycled through the environment. This research has led to an understanding that there is global feedback between ecosystems and the physical parameters of this planet, including minerals, soil, pH, ions, water, and atmospheric gases. Six major elements (hydrogen, carbon, nitrogen, oxygen, sulfur, and phosphorus; H, C, N, O, S, and P) form the constitution of all biological macromolecules and feed into the Earth's geochemical processes. From the smallest scale of biology, the combined effect of billions upon billions of ecological processes amplify and ultimately regulate the biogeochemical cycles of the Earth. Understanding the relations and cycles mediated between these elements and their ecological pathways has significant bearing toward understanding global biogeochemistry.

The ecology of global carbon budgets gives one example of the linkage between biodiversity and biogeochemistry. It is estimated that the Earth's oceans hold 40,000 gigatonnes (Gt) of carbon, that vegetation and soil hold 2070 Gt, and that fossil fuel emissions are 6.3 Gt carbon per year. There have been major restructurings in these global carbon budgets during the Earth's history, regulated to a large extent by the ecology of the land. For example, through the early-mid Eocene volcanic outgassing, the oxidation of methane stored in wetlands, and seafloor gases increased atmospheric CO (carbon dioxide) concentrations to levels as high as 3500 ppm.

In the Oligocene, from twenty-five to thirty-two million years ago, there was another significant restructuring of the global carbon cycle as grasses evolved a new mechanism of photosynthesis, C photosynthesis, and expanded their ranges. This new pathway evolved in response to the drop in atmospheric CO concentrations below 550 ppm. The relative abundance and distribution of biodiversity alters the dynamics between organisms and their environment such that ecosystems can be both cause and effect in relation to climate change. Human-driven modifications to the planet's ecosystems (e.g., disturbance, biodiversity loss, agriculture) contributes to rising atmospheric greenhouse gas levels. Transformation of the global carbon cycle in the next century is projected to raise planetary temperatures, lead to more extreme fluctuations in weather, alter species distributions, and increase extinction rates. The effect of global warming is already being registered in melting glaciers, melting mountain ice caps, and rising sea levels. Consequently, species distributions are changing along waterfronts and in continental areas where migration patterns and breeding grounds are tracking the prevailing shifts in climate. Large sections of permafrost are also melting to create a new mosaic of flooded areas having increased rates of soil decomposition activity that raises methane (CH) emissions. There is concern over increases in atmospheric methane in the context of the global carbon cycle, because methane is a greenhouse gas that is 23 times more effective at absorbing long-wave radiation than CO on a 100-year time scale. Hence, there is a relationship between global warming, decomposition and respiration in soils and wetlands producing significant climate feedbacks and globally altered biogeochemical cycles.

7

Soil Science

Soil plays a vital role in sustaining life on the planet. Nearly all of the food that humans consume, except for what is harvested from marine environments, is grown in the Earth's soils. Other obvious functions that soils provide humans include fiber for paper and clothing, fuelwood production, and foundations for roads and buildings. Less obvious functions that soils serve are providing a medium to attenuate pollutants and excess water, groundwater recharge, nutrient cycling, and habitat for microorganisms and biota. Soils also have many secondary uses such as ingredients in confectionaries, insecticides, inks, paints, makeup, and medicines; uses of clays range from drilling muds, pottery, and artwork, to providing glossy finishes on various paper products.

Soil is a critical component of nearly every ecosystem, but is often taken for granted. Soil can be thought of as the ecosystem foundation, as soil productivity determines what an ecosystem will look like in terms of the plant and animal life it can support. For example, in forest ecosystems, soils can determine species composition, timber productivity, and wildlife habitat, richness, and diversity. The role soil plays in forests is also critical to maintaining water quality and long-term site productivity. In cultivated fields, soil quality plays a significant role in crop productivity since soil nutrients and soil physical properties can directly impact yields. In urbanized areas, soil plays a vital role in reducing runoff through infiltration and nutrient attenuation. The value of soil is easily overlooked until soil quality becomes degraded and the critical services the soil once provided are diminished.

SOIL OVERVIEW

Soil Defined

The definition of soil is relative to the function it provides to the person(s) defining it. From a morphological stance, the Natural Resource

Conservation Service (NRCS) defines soil as: "a natural body comprised of solids (minerals and organic matter), liquid, and gases that occurs on the land surface, occupies space, and is characterized by one or both of the following: horizons, or layers, that are distinguishable from the initial material as a result of additions, losses, transfers, and transformations of energy and matter or the ability to support rooted plants in a natural environment" (Soil Survey Staff 2014a). The Soil Science Society of America (SSSA) defines soil in terms of its genetic and environmental factors: Soil is "[T]he unconsolidated mineral or organic matter on the surface of the Earth that has been subjected to and shows effects of genetic and environmental factors of: climate (including water and temperature effects), and macro- and microorganisms, conditioned by relief, acting on parent material over a period of time. A product-soil differs from the material from which it is derived in many physical, chemical, biological, and morphological properties and characteristics" (SSSA 2008).

The Soil Profile

The vertical section of soil that shows the presence of distinct horizontal layers is known as the soil profile (SSSA 2008). The term horizon refers to the individual or distinct layers within the soil profile. Most soils are composed of several horizons. Typically, horizons of a soil profile will follow the topography of a landscape. Designation of horizon boundaries also comes from measurements of soil color, texture, structure, consistence, root distribution, effervescence, rock fragments, and reactivity.

The uppermost layer, the O horizon, consists primarily of organic material. Forested areas usually have a distinct O horizon. However, in some settings such as a grassland or cultivated field, there may be no O horizon present. Factors such as erosion or constant tillage contribute to the lack of organic matter. The O horizon has three major sub-classifications, or subordinate distinctions (designated by the lowercase letter): hemic (Oe), fibric (Oi), and sapric (Oa). The hemic layer consists of decaying material that is slightly decomposed, yet the origin is still identifiable. The fibric layer is composed of organic material that is slightly more decomposed and unidentifiable, but is not decayed entirely. The sapric layer consists of fully decomposed material whose origin is completely unidentifiable.

The A horizon is a mineral horizon that is formed at or just below the soil surface. It is commonly referred to as the "surface soil." Some characteristics of an A horizon may include the accumulation of organic matter and/or the presence of a plow pan. A plow plan (or plow layer) is a common characteristic of soils that have undergone conventional tillage at some point in recent time. The darkness of the A horizon can sometimes be attributed to the movement of organic matter from the overlying O horizon. Soils under intense cultivation will incorporate materials that would normally be considered part of the O horizon. These organic materials also contribute to the A horizon leading to a higher organic content than other horizons.

The E horizon (eluvial layer) is a common mineral horizon in forest soils that is distinguished by its lack of clay, iron (Fe), or aluminum (Al). The loss of the above materials is known as eluviation, which entails that these substances and dark minerals have been stripped from the soil particles. Clay, Fe, and/or Al are removed from the E horizon via leaching, which causes its light color compared to the adjacent horizons. Leaching is the loss of nutrients from the root zone due to the movement of water through the soil profile. The E horizon is comprised of concentrations of quartz, silica, or other minerals that are less susceptible to leaching.

The B horizon, known as the "zone of accumulation", occurs below the O, A, and/ or E horizons, if present. The B horizon receives deposits of illuviated materials such as clay particles, Fe and Al oxides, humus (organic matter formed from the decay of plant and animal matter), carbonates, gypsum, and silicates leached from the overlying horizons. The common presence of Fe and Al oxide coatings often give the B horizon a redder or darker color than the adjacent horizons.

The C horizon is the soil layer that generally sees little influence from pedogenic weathering processes and is therefore comprised of partially weathered parent material. The C horizon represents a transition between soil and bedrock. As the upper portion of the C horizon undergoes weathering, it may eventually become part of the overlaying horizons. There is an obvious shift in soil structure between strongly developed B and C horizons that aids in identifying the horizon boundary in the field; however, the structure shift may be more subtle in weakly developed soils.

Under the C horizon comes the R horizon, or bedrock. Depending on the geographic location, environmental conditions, and landscape position, bedrock may be found in excess of 100 feet deep or merely centimeters from the soil surface. Bedrock is a consolidated layer of rock material that gave way to the soil properties found on the site. Bedrock is occasionally disrupted or broken up by tree roots, but roots generally cannot cause enough stress on the rock to fracture it, so much of the deeper bedrock weathering is biochemical in nature. The layer of freshly weathered material, in contrast to the solid rock (i.e., bedrock), is generally termed saprolite/saprock.

The division of the soil profile termed the regolith is defined as "the unconsolidated mantle of weathered rock and soil material on the Earth's . surface which extends from the soil surface to the bottom of the parent material" (SSSA 2008). So basically, the regolith is the heterogeneous material that lies on top of solid rock. The soil solum is the weathered soil material in the upper soil horizons (typically A, E, and B horizons) located above the parent material (C horizon). Not every soil profile is comprised of the same horizons. Some profiles will contain O, A, E, B, C, and R horizons while another soil profile may only be composed of a C and R horizon. These differences in horizonation are what make soils unique. The unique characteristics of soil allow soil scientists to classify soils into different categories via Soil Taxonomy (see Soil Classification section).

Soil Formation

The 5 soil-forming factors that influence the development of soil were first termed by Hans Jenny, an American soil pedologist in the early- to mid-1900s. The factors he determined as essential in the formation of soil include: parent material, climate, biota, topography, and time (Jenny 1994). Hans presented these factors as a formula: s = f (PM, Cl, O, R, T). The formula roughly translates to soil is a function of parent material, climate, organisms, relief, and time. This section will briefly discuss the role of each of the soil-forming factors and how each factor aids in soil development.

Parent material

Parent material is the unconsolidated and chemically-weathered mineral or organic matter from which the soil is developed. It consists of any number of combinations such as limestone, sandstone, or even volcanic rock. The area of the soil profile known as the "C Horizon" is comprised of parent material. Parent materials attribute to both the chemical and physical properties of a soil and can affect soil drainage, porosity, and plant available water, among other things. Parent material in any location can vary greatly even in areas adjacent to one another. For instance, Illinois contains eight different soil regions that are a result of glaciations, wind-blown materials, and continuous weathering of parent material.

Climate

In early soil profile development, parent material gives way to the major physical and chemical properties of a soil. As the soil becomes more developed (through horizon formation and increased soil structure over time), its characteristics depend more heavily on the climatic conditions to which it is exposed. Climate plays a significant role in the formation of soil and any number of climate-related occurrences (e.g., precipitation and temperature) may influence soil development.

Biota

Soil profile features related to biotic (plant and animal) activity such as burrows, mounds, root channels, and worm castings contribute to soil profile development because each of these processes change the porosity of the soil. The burrowing of animals, much like old root channels, creates large pores for rapid movement of water, gases, and solutes through the soil. The structure of some surface horizons is formed entirely by animal activity (earthworms, ants, termites, and other organisms). Earthworms are capable of consuming their own body weight in food daily (Minnich 1977). They are also responsible for the 'sinking' of objects through the soil profile over time (Darwin 1897). Charles Darwin devoted his last book, "The Formation of Vegetable Mould Through the Action of Worms" to the process of bioturbation, the process in which plants and animals facilitate the mixing, or rearrangement, of the soil profile.

Much like animals, plants can have a strong influence on soil properties. For example, whether a soil is formed under forest cover or prairie vegetation can greatly influence the carbon inputs into the system and ultimately how the soil is classified (see Soil Classification section). Plant roots increase infiltration, break up dense soil layers, and can pull nutrients and moisture from deep within the soil profile. Further, windthrow (uprooting of trees by wind) can create a pit and mound topography that helps physically weather the soil by dislodging and breaking rock. Windthrow can also expose deeper soils to surface conditions that may lead to accelerated chemical weathering.

Topography

Topography (slope and aspect) also has a strong influence on soil characteristics. Slope can influence erosion and deposition along a hillslope. Steep soils are susceptible to accelerated erosion and generally have a shallower A horizon and overall less development. Conversely, soil development in flat areas is heavily influenced by soil drainage, where well-drained soils tend to have greater development compared to poorly drained soils. A flat, poorly-drained soil retains water, leaving the soil profile saturated, which slows soil profile development. In well-drained soils, E horizons can develop above a well-developed B horizon due to the eluviation (transport of materials via water) of clay from the upper soil horizons. Additionally, the aspect (or horizontal direction) on which a soil is formed affects profile development. South-facing slopes receive more intense solar radiation than north-facing slopes, which affects soil temperature and moisture conditions. In the Appalachians, south-facing slopes are dominated by coniferous species while north-facing slopes are dominated by hardwood species due to differences in microclimate. Furthermore, topographic position (i.e., summit, shoulder, backslope, and toeslope) influences the development of soils. Generally, summits are well-drained and have strong horizon development, whereas shoulder and backslopes are influenced by erosion and have shallower topsoils and less infiltration. Toeslopes typically accumulate organic matter and topsoil from both alluvial (deposited by streams) and colluvial (primarily deposited by gravity) inputs, leading to thicker A-horizons overlaying relatively young soils.

Time

The duration of time a soil has undergone development is determined largely by the degree of weathering of the soil. Time has two separate meanings in terms of soil formation. Soil is influenced by both chronological and physiological time. The 'age' of a soil commonly refers to the degree or amount of weathering the soil has undergone. This age is not referring to chronological time, but the physiological characteristics attributed to the soil via the weathering process.

SOIL CLASSIFICATION

Around the globe, many soil classifications systems have been developed to categorize soils into groups based on morphological and/or chemical properties. The most widely-used classification system is the Soil Taxonomy system that was made known by the United States Department of Agriculture (USDA) (Soil Survey Staff 1999). This system is a morphogenetic system that utilizes both quantitative factors and soil genesis themes and assumptions to guide soil groupings (Buol et al. 1997). "Keys to Soil Taxonomy", a free publication distributed by the USDA, is a great resource for in-depth classification of soils (Soil Survey Staff 2014a). The USDA has also published a free resource titled "Illustrated Guide to Soil Taxonomy" that is similar to "Keys of Soil Taxonomy", but written for a broader audience (Soil Survey S taff 2014b). Both resources can be found on the USDA website listed in the Soil Resources section of this article.

The Soil Taxonomy system is a hierarchical scheme consisting of 6 classification levels. In order from broadest to narrowest, the levels of classification are: 1) Order, 2) Suborder, 3) Great Group, 4) Subgroup, 5) Family, and 6) Series. Currently, there are 12 soil orders, 65 suborders, 344 great groups, <" 18,000 subgroups, and over 23,000 soil series (Bockheim et al. 2014).. The defining characteristics of the broadest levels of classification are based on soil-forming processes and parent materials, whereas the narrower levels become much more specific and consider the arrangement of horizons, colors, textures, etc. The soil-forming factor "climate" has a predominate role in Soil Taxonomy classification, followed by parent material, and biota; topography and time are not utilized in defining taxa (Bockheim et al. 2014).

Soil Orders

Entisols are the 'youngest', or most recently formed soils of all the soil orders. Characteristics of Entisols include weak profile development where very little, if any, horizonation can be documented. Entisols sometimes contain a weakly formed A or Ap (plow layer) horizon. These soils can be found on steep slopes with severe erosion, on floodplains that receive alluvial deposits, and any number of scenarios in between.

Another soil order with notably weak profile characteristics is the Inceptisol order. When Soil Taxonomy was first established in 1975, Inceptisols were commonly referred to as the 'wastebasket soil order '. These soils generally did not fit into the other soil orders at the time. When additional soil orders were introduced, many Inceptisols were reclassified and the 'wastebasket' title no longer applies. Inceptisols are somewhere between the stages of no profile development and weak profile development. Inceptisols are commonly found along major rivers and streams due to the weak profile development. Over time, both Entisols and Inceptisols have the potential to develop more horizons, at which time they would likely be reclassified into a different soil order. The features of 'young' soils like Entisols and Inceptisols are more heavily influenced by their parent material, whereas 'old' soils are more influenced by climate and vegetation factors.

Gelisols are also 'young' soils in regard to geologic time and were developed under cold temperatures or frozen conditions. These soils are often associated with permafrost conditions and cryoturbation (frost churning) in places like Canada and Alaska. Permafrost is a perennially frozen soil horizon (SSSA 2008).

Mollisols are commonly referred to as the 'prairie soil'. These soils were formed primarily under grassy prairies and are characterized by their high organic matter content, dark color, and deep A horizon. The A horizon must be greater than 8 inches in depth and requires at least a 50 % base saturation (at least 50 % of the cation exchange sites are occupied; see Soil Chemistry section for more information). Mollisols are common in the midwestern United States where native prairies once dominated the landscape.

Alfisols, formed under deciduous forests, are also very common in the midwestern United States. Alfisols are generally found in humid regions of the world and often contain an E horizon in the soil profile. These soils must have a base saturation of at least 35 %.

Spodosols generally originate from coarse-textured (i.e., increased sand content), acidic parent materials. Spodosols are formed under forest vegetation, especially coniferous forests due to the buildup of pine needles that inherently have high acidic resins. When pine litter decomposes, strongly acidic compounds are leached through the coarse materials, transporting Fe, Al, and humus (Brady and Weil 2007). Thus, an illuvial layer of humus and Fe/Al oxides form. Like the Alfisols, an E horizon is commonly found in Spodosol soil profiles. In many cases, a Spodosol will have a white E horizon on top of a bright red B horizon. Spodosols are associated with loamy or sandy soil conditions and can be found in Wisconsin, Michigan, the northeastern United States, and on the coastal plains of the eastern and southeastern United States.

Aridisols are commonly associated with semi-arid and arid regions. These regions have a low mean annual rainfall. The lack of moisture in the soil affects the soil development and weathering process. Therefore, these soils are primarily affected by physical weathering, not chemical weathering (weathering processes discussed in the Soil Chemistry section). Aridisols are characterized by a high base saturation percentage of <" 100 %. These soils can be found throughout the deserts or drier areas of the western United States.

Ultisols can be found in humid and warm regions such as the southeastern United States. Ultisols have a high amount of clay mineral weathering and translocation that leads to a subsurface accumulation of clays (Brady and Weil 2007). They have a base saturation of < 35 % and are naturally less fertile than Alfisols or Mollisols. However, Ultisols respond favorably to nutrient management and are cultivated in many regions of the world. These soils are characterized by a high degree of weathering and are typically more acidic than Alfisols, but less acidic than Spodosols.

Oxisols are the most highly weathered soil order in the U.S. classification system. Oxisols get their name from being oxidized. They are dominated by high clay content and Fe/Al hydrous oxides that typically give the soil a red hue. Oxisols can be found in tropical and sub-tropical regions of the world such as Hawaii, Puerto Rico, South America and Africa. Oxisols are generally formed in wetter environments, but can be found in areas that are presently drier than during the time the soils were formed (Brady and Weil 2007).

Vertisols are soils that lack profile development due to the expansion and contraction of clay-rich soil. These processes cause the soil to mix, which does not allow for clear soil profile development. During dry conditions, the soil shrinks and cracks form that can extend to <" 80 cm deep and 2 to 3 cm wide. Vertisols are found in the southern United States in places like southeast Texas and eastern Mississippi where smectite, montmorillonite, and vermiculite clays are present.

The Andisol order was developed in 1990. Previously, these soils were grouped with the Inceptisol order. Andisols are soils of recent origin (young) that are developed from volcanic materials. They are found in Hawaii and the northwestern United States in places like Washington, Idaho, and Oregon.

Histosols are the only organic soil order in the classification system. Histosols are comprised of several different subhorizons within the O horizon and contain at least 20 % organic matter. Histosols occupy a small total area, but are found in various places in the United States and Canada such as Wisconsin, Minnesota, the Florida Everglades, and along the Gulf Coast.

SUBORDER, GREAT GROUP, SUBGROUP, FAMILY, AND SERIES

Suborders categorize properties associated with a climatic connotation of the soil. Great groups account for the most significant properties of the soil as a whole, including the type and arrangement of soil horizons, temperature regimes, and moisture regimes. Scientists use subgroups to further classify soils by assessing the degree of similarity between particular soils and grouping them accordingly. These are intergrades that reflect transitions to other orders, suborders, or great groups. The family grouping has similar physical, chemical, and mineralogical properties, which often relate to plant growth. A soil series is the lowest level of taxonomy, or the most specific to the soil in question. The soil series classification narrows characteristics of similar soils down to a local level where not only physical, chemical, and mineralogical properties matter, but also management, land-use history, vegetation, topography, and landscape position. Most soil series are named based on the location where the series was first discovered. An example of the taxonomic classification for the Illinois state soil is provided (Table 1).

Table 1. Example of the taxonomic classification according to U.S. Taxonomy for the state soil of Illinois, USA.

Taxonomic Level	Example for IL State Soil
Order	Mollisol
Suborder	aquoll
Great Group	Endoaquoll
Subgroup	Typic Endoaquoll
Family	Fine-silty, mixed, superactive, mesic Typic Endoaquoll
Series	Drummer

Pedologists (scientists who study the origin and composition of soils as a component of natural systems) continue to learn more about soil morphology, chemistry, mineralogy, and biota, thus tighter limits on groupings of soils are being defined. Soil scientists are constantly reevaluating soil taxonomy as technology and science progress. It is important to remember that soils are dynamic, living environments. Soils are ever-changing through both chemical and physical weathering processes and can vary across the landscape, resulting in corresponding changes in soil classification over time.

Soil Physical Properties

In this section, soil physical properties will be introduced, examples for measurement will be provided, and the applicability in the field will be discussed. Soil physical properties have a profound effect on how soils influence soil quality and productivity. Oftentimes soil quality is driven by soil physical properties that determine nutrient and moisture levels in soils. The physical properties of soil include soil texture, bulk density, water holding capacity, organic matter content, soil structure, soil color, and soil consistence. A useful field guide for describing soil properties is "Field book for describing and sampling soils" (Schoeneberger et al. 2012).

Soil Texture

Soil texture is determined by the amount of sand, silt, and clay in a soil sample. Table 2 shows the size comparison of sand, silt, and clay particles based on the United States Department of Agriculture (USDA) System, one of the most commonly used among several particle size classification systems. Clays have a particle size diameter of < 0.002 mm, silts between 0.002 and 0.05 mm, and sands 0.05 to 2.0 mm. Percentages of sand, silt, and clay categorize soil into different textural classes. Particle size percentages can be compared to the USDA soil textural triangle to determine soil texture. For instance, a soil with 18 % clay, 60 % sand, and 22 % silt would be considered a sandy loam. Soil texture can be determined in the lab via particle-size analysis using the hydrometer method; this method determines particle-size distribution by measuring the time it takes for soil particles in water to settle out of suspension. The hydrometer method can be time consuming, requires certain equipment, and is not optimal for most people. However, anyone can determine textural class using the "Feel" Method. This involves forming a moist soil sample into a ball and squeezing it between the thumb and index finger to make a ribbon. Texture is determined by ribbon length (if one can be formed at all) and grittiness or smoothness of the sample. A step-by-step guide to the "Feel" Method is available on the NRCS website (see Soil Resources section). With practice, the "Feel" Method provides an immediate and accurate way to determine textural class, and even percent sand, silt, or clay, while in the field. Laser Particle Size Analysis (LPSA) has become more commonly used in labs across the country. Although this method is accurate, it requires training/skill to use the analytical instrument and the initial cost is high.

Table 2. Particle diameter (mm) of sand, silt, and clay based on the United States Department of Agriculture System.

Texture Size Class	Particle Diameter (mm)
Very coarse sand (vcos)	1.0–2.0
Coarse sand (cos)	0.5–1.0
Medium sand (s)	0.25–0.5
Fine sand (fs)	0.1–0.25
Very fine sand (vfs)	0.05–0.1
Silt (si)	0.002–0.05
Clay (c)	< 0.002

Bulk Density

Bulk density is the mass of the soil in relation to a known volume of soil and is often used as an indicator of soil compaction. Soil compaction decreases the ability of a plant's roots to penetrate through the soil profile. Bulk density is related to soil textural class and soil porosity. Soils containing a high percentage of porosity will have a lower bulk density. Typically fine-textured soils have lower bulk densities than coarse-textured soils because of increased pore space both between (interstructural pores) and within (matrix pores) soil aggregates. Ideal bulk densities for plant growth range from < 1.10 g cm for clays to < 1.6 g cm for sands (Table 3). Bulk density is also affected by grain size and arrangement of coarse-textured soils. If sandy soils are loosely-packed and of uniform size, porosity is higher and therefore bulk density is lower than sandy soils with tightly-packed aggregates of different grain sizes. If a textural class has a bulk density reading that exceeds the growth-limiting factor for that class, then conditions for plant life are not optimal. Daddow and Warrington (1983) determined growth-limiting bulk densities according to particle size. Generally, the greater the clay percentage is within a soil, the lower the growth-limiting bulk density of the soil because of the increased pore space in clays. The growth-limiting bulk density is <" 1.40 g cm for soils with > 80 % clay and <" 1.70 g cm for soils with < 20 % clay (Daddow and Warrington 1983). The impact of bulk density on root growth according to textural class is shown in Table 3. In addition to texture, land cover and management also impact bulk density. Typically, bulk density readings in a forest are much lower than readings in an agricultural field or an urban area. Forest soils tend to have a higher amount of porosity because of tree rooting, increased biotic activity, increased organic matter on the soil surface, and less anthropogenic disturbances. Bulk density readings, soil texture data, infiltration capacity readings, and penetration resistance data can be compiled to explain soil porosity or degree of compaction. Compaction affects both root penetration and the soil's ability to allow precipitation to infiltrate into the soil profile. Bulk density is commonly measured using a core method where a known volume of soil is collected with a cylindrical soil core, then oven-dried, and weighed. Results are reported as a mass per unit volume (e.g., g cm). Bulk density can also be determined via the clod method, where a clod (or ped) of soil is coated in paraffin wax and placed in water to determine the exact volume based on displacement. The clod is then oven-dried, weighed, and reported on a mass/volume basis like the core method. 3-D scanners are also used to determine accurate clod volumes.

Table 3. Relationship between bulk density and root growth based on soil texture. Source: Arshad et al. 1996.

Textural class	Ideal bulk density (gcm)	Bulk density that may affect root growth (gcm)	Bulk density that restricts root growth (gcm)
Sands, loamy sands	< 1.60	1.69	> 1.80
Sandy loams, loams	< 1.40	1.63	> 1.80
Sandy clay loams, clay loams	< 1.40	1.60	> 1.75
Silts, silt loams	< 1.30	1.60	> 1.75
Silt loams, silty clay loams	< 1.40	1.55	> 1.65
Sandy clays, silty clays, some clay loams (35 – 45 % clay)	< 1.10	1.49	> 1.58
Clays (> 45 % clay)	< 1.10	1.39	> 1.47

WATER HOLDING CAPACITY

Water holding capacity is the amount of water held by soil. The water holding capacity of a soil is reliant on soil texture, structure, organic matter content, and arrangement of soil pores. Organic matter has a high degree of microporosity, which allows it to retain more water. Therefore, soils with a higher amount of organic matter and/or a large percentage of micropores (e.g., fine-textured soils like clays) generally have a higher water holding capacity. Soil compaction also impacts water holding capacity since soil compaction weakens soil structure and collapses pores, thereby decreasing the soil's ability to hold water. There are several different terms used to discuss water capacity in soils. Available water is the amount of water that is readily available for plant uptake; conversely, unavailable water is water that plants cannot utilize. Gravitational water (water that drains freely until field capacity is reached) is a form of unavailable water. Field capacity refers to the water held by the soil following 24 to 48 hours of free drainage and is available for plant uptake. The permanent wilting point is the point at which plants wilt and are unable to recover due to a lack of plant available water. Lastly, saturation occurs when all pores are filled with water .

Soil Structure

Soil structure is the description of how individual soil particles (sand, silt, and clay) are arranged into soil aggregates (also called peds) and reflects both physical and chemical weathering. Several factors influence soil structure, including soil texture, soil moisture, organic matter content, compaction, the activity of soil organisms, and management practices. Soil structure is easiest to observe in dry soil. When characterizing soil structure, the shape, size, and grade of the structural units (peds) are defined. The primary soil structure shapes are granular, platy, blocky, prismatic, and columnar. Granular soil structure contains peds that are generally small and round and are commonly found in horizons near the surface where high amounts of root activity are present and porosity is greater. Platy soil structure consists of peds that are plate-like (flat and thin) and usually oriented horizontally. Platy soil structure may occur throughout the soil profile, but is common in E horizons or compacted soil. Blocky soil structure consists of sharp-edged peds arranged in square or angular blocks and is typically found in the subsoil, especially in humid regions. Prismatic soil structure contains peds that are longer vertically than horizontally and have flat tops. Columnar soil structure is similar to prismatic, but peds have distinct rounded tops. Both prismatic and columnar soil structure are commonly found in the subsoil of arid and semi-arid regions. Soils can also be defined as "structureless" where soil structure is single-grained (typically sand particles that do not stick together) or massive(aggregation is not present). Aggregate size is typically described as very fine, fine, medium, coarse, or very coarse and the size limits of these classes depend on the aggregate shape. Grade describes the distinctness of the peds and is classified as weak, moderate, or strong. More information on size and grade can be found in the Soil Survey Manual (Soil Survey Division Staff 1993). A soil structure description is written as "grade, size, shape", such as "moderate medium subangular blocky structure." Soil structure greatly impacts soil porosity, thereby influencing soil water movement.

Soil Consistence

Soil consistence is a measure of the response of soil to applied pressure at various moisture contents. In other words, how well does the soil hold together when under applied stress? Soil consistence is measured separately for dry, moist, and wet soil because moisture content impacts how a soil responds to pressure. In the field, soil consistence is measured by testing the ease with which the soil is crushed between the thumb and forefinger, or underfoot. Additionally, the terms stickiness and plasticity are often used to describe wet consistence. Stickiness refers to how well wet soil adheres to other objects (like your fingers) after pressure is released, while plasticity describes the malleability of wet soil. Further reading on the topic, including the scale used to describe consistence at different moisture contents, can be found in chapter 3 of the Soil Survey Manual (Soil Survey Division Staff 1993). Soil consistence is one of many soil physical properties used to determine site suitability for agriculture or engineering purposes.

Soil Color

Soil color is important in determining soil classification. The Munsell color chart, a book containing standard color chips similar to paint chips found at a hardware store, was developed as the standard system for determining soil color. Color is determined from three characteristics: hue, value, and chroma. Hue refers to the degree of redness or yellowness of soil. Value refers to the lightness or darkness of soil. Chroma refers to the brightness or dullness of soil. A freshly exposed face or ped is used to determine color. Using moist soil is most common and dry soil may be misted with water. In direct sunlight, the moist soil ped is compared to the Munsell color chart to determine the hue, value, and chroma. Soil color varies with topography, climatic factors, and soil depth, among other variables. Drainage characteristics of a soil can have a large impact on soil color; thus, soil color can reveal insights into the local hydrologic regime. Soils that are well-drained tend to be brighter than poorly-drained soils. Poorly-drained soils create anaerobic conditions in which the Fe in the soil is reduced, resulting in very dull colors. Soils with extremely reduced conditions and a chroma < 2 are referred to as "gleyed" soils.

Organic Matter

Organic matter content has a profound influence on both soil processes and soil quality. In the field, high organic content may be recognized by a dark soil color in the surface horizons. In the lab, organic matter content is quantified by Loss on Ignition, a process in which a soil sample is exposed to high temperatures ($360°$ C) and the amount of weight lost after exposure is assumed to be organic carbon. Organic matter in soils promotes biotic growth since it serves as a food source for earthworms and other organisms. Organic matter has a high water infiltration capacity, a high moisture holding capacity (it can hold between 80 and 90 % of its weight in water), and contains many plant-essential nutrients. The increased water holding capacity allows more water to be available to plants over a longer timeframe.

Along with aiding in the retention of soil moisture, organic matter protects the soil against the kinetic energy of raindrops and also acts as an insulation layer for the soil surface. Without organic matter, bare mineral soil is much more susceptible to accelerated erosion processes. Since billons of tons of soil in the world are displaced every year, it is important for soil organic matter to remain intact whenever possible. Additionally, organic matter acts as a binding agent for nutrients and potential contaminants, and therefore aids in reducing inputs of these contaminants to waterbodies. Depending on environmental conditions, organic matter can be stored in the soil for long periods of time. Warm, humid conditions promote breakdown of organic matter by microorganisms, whereas cooler, drier climates limit decomposition and soils act as a carbon reservoir. Tillage also impacts organic matter storage in agricultural soils. Tillage generally reduces organic matter content, as it improves conditions for decomposition by increasing pore space and moisture content and by exposing organic matter adsorbed to soil aggregates; converting to no-till agriculture has been shown to increase organic matter content in soils

SOIL WATER MOVEMENT

Soil water is the amount of water present in the vadose zone, or the zone of unsaturated flow, of the soil profile. The term groundwater refers to the area of saturated flow in the soil. Water enters the soil profile through the process of infiltration, and then moves through the soil profile via percolation. These processes depend on various soil properties that range from soil porosity to the shape and arrangement of soil peds. Water moves through the available pore space more readily and at a faster rate in soils with granular structure when compared to the longer flow path that platy soil structure provides. The percentages of micropores, mesopores, and macropores in the soil horizons also influence how quickly water enters into and moves through the soil profile.

Micropores, or capillary pores, are the smallest soil pores and contain most of the plant-available water in soil. Water is held in these pores by the combined matric forces of adhesion and cohesion. Adhesion is the ability of a water molecule to adhere to a soil particle. Cohesion is the ability of a water molecule to stick to itself (or other water molecules). Mesopores are medium-sized pores formed by numerous processes like the shrinking and swelling of clays, earthworm activity, and freeze-thaw cycles. These pores are primarily drained during the period of free drainage (gravity-driven drainage immediately following a rain event), although some water drained by mesopores is eventually plant available if the drainage network leads into micropore spaces. Macropores are the largest of the soil pore classification. Some macropores are formed by the burrowing of animals (non-matrix pores) while others exist because tree roots that once occupied the soil have rotted away. Drainage in macropores is gravity-driven and rapid. The water in these pores is quickly replaced by air; therefore, macropores cannot supply plants with needed water. Soils containing a range of pore sizes will have good drainage and air available to plants (from large pores), in addition to water plants can access (from small pores). Ideal pore distribution is generally found in a well-structured soil with the majority of small pores within soil aggregates (matrix pores) and large pores between soil aggregates (interstructural pores) (Kimmins 1997). Forest soils contain a larger percentage of organic matter that results in better soil aggregation and porosity; thus, moisture availability in forests is usually greater than in agricultural or urban environments.

Soil Water Potential

Soil water potential is the measure of the energy status of the soil water and reflects the amount of water available for plant uptake. Water moves from areas of high energy to areas of low energy within the soil.

Soil water has three types of energy: gravitational, matric, and osmotic. Gravitational potential refers to the energy related to the elevation of water within the soil; water at the soil surface has a high gravitational potential, but once it reaches the water table the gravitational potential is zero. Matric potential is the energy associated with adhesion and cohesion within the soil matrix (or profile) and is a negative potential since it opposes gravitational and osmotic potential. Osmotic potential measures dissolved chemicals within the soil; the higher the concentration of dissolved chemicals, the less room water has to move, though typically this has little contribution to overall soil water movement except in saline soils. Soil water potential is presented as the following simplified formula:

where - is total soil water potential, - is gravitational potential, - is matric potential, and - is osmotic potential. This energy status reflects the amount of energy required to extract water from the soil.

Soil water potential is typically measured in bars or kilopascals of pressure. It is important to note that soil water potentials in the vadose zone are recorded as negative pressures. Positive pressure values would indicate that the soil is saturated and when this occurs, it means that the water being evaluated is not true vadose zone soil water since the vadose zone is unsaturated. The most common device used for measuring soil water potential in the field is a water-filled tube called a tensiometer. Water flows from a higher potential (inside the tube) to a lower potential (within the soil matrix) until an equilibrium is reached and the soil water potential is displayed as a negative value (range of 0 to "85 kPa) on the tensiometer vacuum gage. The closer the value is to 0, the closer the soil is to saturation. Smaller values (closer to "85 kPa on the negative scale) mean drier soil and that more energy is required to extract water.

Texture and structure may vary from one horizon to another, which can influence the movement of water through the soil profile. Soil texture affects how water moves through the soil and how much water can be stored in the soil. Soil structure affects the ability of roots to penetrate the soil, the amount of water a plant can uptake, and water movement through the soil. A poorly structured soil may have as little as 35 % total porosity, while a well-structured soil of the same texture may have 65 % total porosity (Kimmins 1997). A well-structured soil will be more efficient at soil water movement and plant uptake. Additionally, because small pores have more matric potential (adhesion and cohesion forces) than large pores, fine-textured soils that contain more micropores generally hold more water than coarse-textured soils containing a greater percentage of macropores. By the time field capacity (FC) (soil is not saturated, but wet) is reached, most of the gravitational water is lost, larger pores become filled with air, and water is held within smaller pores. At FC, most water is plant available and soil water potential values typically show <" "10 to "30 kPa (Brady and Weil 2007). The permanent wilting point (PWP) is the potential at which soil water is unavailable for plant uptake (<" "1500 kPa). At PWP, plants cannot recover from water stress and will remain wilted. However, there is still a small amount of hygroscopic water in the smallest pores. The available water content (AWC) is considered to be the soil water retained between FC and PWP and is presented as the formula:

From this formula, the total available water content (TAW) can be obtained by multiplying the AWC by the depth of the plant root zone (Rd) or:

SOIL CHEMISTRY AND PLANT UPTAKE

Soil chemistry plays a key role in vegetative productivity and species composition and is largely determined by weathering of rock, rock type, the cation exchange capacity of the soil, acid production resulting from microbial and root respiration, and management strategies of the soil. The soil provides nutrients necessary for plant growth. The sources of plant nutrients range from biogeochemical cycling inputs, decaying organic matter, amendments added by humans (e.g., fertilizers and pesticides), and nutrients naturally occurring within mineral soil.

Weathering

There are two processes associated with soil weathering: physical and chemical weathering. Physical weathering is the establishment of sufficient stress on a rock so that it physicallybreaks the rock. Mechanisms that fall within the category of physical weathering include erosion, freeze-thaw cycles, bioturbation (plants and animals physically disturb the soil and plant roots may physically crack rock apart), and formation of cracks or gaps in soils. Parent material weathers to release primary minerals to soils.

Chemical weathering is defined as a change in the chemical nature of rock, resulting in a change in mineral structure, reactivity, surface area, and particle size; it is driven by the instability of primary minerals at the earth's surface. The minerals must be exposed to the environment for weathering to take place. Primary minerals include quartz, feldspars, muscovite, and biotite. These minerals have persisted through geologic time and their mineral composition shows only negligible differences from their original state. Chemical weathering is generally responsible for the formation of secondary minerals, which are derivatives of primary minerals. Secondary minerals are the fine materials that make up clay particles in the soil. The presence of clay in a soil signifies an active history of weathering. An example of a primary mineral weathering to form a secondary mineral is the breakdown of feldspar (primary) into clay (secondary). Chemical weathering reactions increase in warm, humid climates and are also enhanced by the presence of water and oxygen, as well as biological agents including the acids produced by microbial and plant-root metabolism (Brady and Weil 2007).

The four primary chemical weathering processes are oxidation, reduction, hydration, and hydrolysis. Oxidation occurs when the oxygen supply is high and an element loses electrons; an example is when ferrous Fe combines with oxygen to form ferric Fe oxide (or rust). Reduction occurs when an element gains electrons and generally occurs in anoxic (oxygen-depleted) environments. Hydration is a result of the association of water molecules onto the mineral structure (e.g., anhydrite and water forms gypsum). Lastly, hydrolysis is essentially an attack on the silicate structure by hydrogen ions, meaning water breaks down the rock. Chemical weathering weakens the rock structure and makes it more susceptible to additional weathering.

Cation Exchange Capacity

Clay and organic matter have what is referred to as cation exchange capacity (CEC), which is the total sum of exchangeable cations that a soil can adsorb. Clay and organic matter are negatively charged, thereby possessing the ability to adsorb and hold cations (positively charged ions) onto what is known as the cation exchange complex. Cations, such as K, Na, and Ca, can be adsorbed onto soil or organic colloids (very small chemically reactive particles with a large surface area per unit mass), making the cations available for plant uptake by preventing cation leaching from the system (Brady and Weil 2007). Soils high in organic matter and/or clay generally have a greater CEC than soils with little humus or clay (i.e., sandy soils), though CEC can vary greatly depending upon the type of clay and amount of organic matter present. CEC must be measured in a lab.

Soil pH

Soil pH is a measure of the hydronium ion in the soil solution, which determines the acidity or alkalinity of the soil. Soil pH varies by region and acidic soils are typically found in wet climates, whereas alkaline soils are generally found in areas with limited rainfall. Soils may become acidic through the weathering of rocks rich in silica, production of acids from organisms, and through the release of acids from decaying organic material. Alkaline soils are a result of the weathering of rocks such as limestone that contain large amounts of calcium carbonate (salts) and from the dust inputs of salts resulting from the evaporation of drainage basins, primarily in arid regions. Soil pH affects nutrient solubility and decomposition rates in soil and thereby has a profound effect on the availability of nutrients to plants. A slightly acidic pH of between 6 and 7 appears to provide optimal nutrient availability to plants, though there are exceptions (Kimmins 1997). Most macronutrients are available within this range. Knowledge of the soil pH profile is especially important in making crop management decisions as some plants may require a set pH range for optimal growth. Soil pH can be measured in the lab using a pH meter or in the field with quick test strips.

Nutrients

There are a number of essential elements required for plant growth that are categorized as macronutrients or micronutrients. The difference lies in the quantities needed for plant growth. Macronutrients are elements that are needed in greater quantities, whereas micronutrientsare required in smaller portions. All are required in varying quantities for optimal growth. Macronutrients include nitrogen (N), potassium (K), calcium (Ca), magnesium (Mg), phosphorus (P), and sulfur (S). Micronutrients include chlorine (Cl), iron (Fe), boron (B), manganese (Mn), zinc (Zn), copper (Cu), molybdenum (Mo), cobalt (Co) and nickel (Ni). Macronutrients can further be classified into primary and secondary nutrients. Generally N, P, and K are considered primary nutrients because they are most often the nutrients limiting plant growth. Ca, Mg, and S are rarely limiting nutrients and as such are considered secondary nutrients. Primary nutrients are discussed in greater detail below.

Nitrogen

Nitrogen is the nutrient needed in the greatest quantities by plants and is generally one of the most limiting to plant growth due to lack of environmental availability. The atmosphere contains a large reservoir of nitrogen gas (<"79 % of the atmosphere), most of which is unavailable to plants and animals. Both inorganic and organic forms of N are found in soils; however, only the inorganic form is available for plant uptake. Most of the soil N (> 95 %) exists in organic form and is therefore unavailable. Nitrate (NO) and ammonium (NH) ions are the two inorganic forms utilized by plants. Nitrate is held primarily in solution and is readily available for plant uptake. Ammonium ions are mostly held on the cation exchange complex. Fortunately, soil microbes can break down organic N (NH) and convert it to forms usable by plants (NH and NO) in a process known as nitrogen mineralization. Immobilization of N can also occur when inorganic N is converted to unusable organic form.

Sources of N are primarily from the atmosphere, biological fixation, and fertilization. Biological fixation accounts for the majority of N inputs in most ecosystems (Kimmins 1997). In agricultural systems, annual biological N fixation is estimated at 50–70 million metric tons of N globally, primarily from the symbiotic relationship between leguminous crops (dominated by soybeans) and rhizobia (nitrogen-fixing bacteria) (Herridge et al. 2008). Nitrogen deficiencies in plants are typically noticed by the yellowing of foliage and stunted growth. Because N deficiencies are widespread and can lead to poor crop yields, N is often applied in excess. Nitrate is extremely mobile and readily leached from the soil if not used by plants. High NOconcentrations (> 10 mg L) in drinking water are a danger to public and ecosystem health, thus nitrate in drinking water is regulated by the Environmental Protection Agency. Additionally, excess N in the soil can lead to incomplete denitrification, which generates nitrous oxide into the atmosphere. Nitrogen is a nutrient requiring careful management, critical for both successful crop production and environmental quality.

Phosphorus

Phosphorus is second only to nitrogen in the amount needed for optimal plant growth. The majority of soil P is derived from mineral weathering and organic matter decomposition, but concentrations of plant-available P are generally very low in soils. Most soil P is in insoluble forms that are unavailable to plants and when soluble forms of P are added via fertilization, they can be converted to insoluble forms over time. Decomposition returns P to the soil, but if crops are removed, the P is also removed from the system. For these reasons, P is often a limiting nutrient in the soil. Phosphorus deficiency in plants is recognized by bluish-green foliage, sparse flowering, and stunted, thin stems. Because plants require large quantities of P and there is little natural available P from weathering, crop harvesting removes the majority of P from the field. Without P amendments, the subsequent crop will likely be P deficient, which has led to the over-fertilization of fields and an accumulation of P in the soil. Unlike nitrate, which is mobile downward through the soil profile, P binds readily to sediment and is easily transported in this bounded state to streams and rivers. Excessive P in runoff can promote the eutrophication (over-abundance of algae growth caused by excessive nutrient accumulation) of downstream water bodies. The resulting algal growth decreases the dissolved oxygen levels in the water, which may lead to substantial fish kills. With the Mississippi River Watershed draining much of the nation's cropland, the Gulf of Mexico consistently has one of the largest dead zones in the world due to eutrophication; the average size of the dead zone over the last 5 years is <" 5,500 mi. For this reason, P management in watersheds is of primary concern.

Potassium

Potassium is available in greater quantities than any other soil macronutrient and is essential in providing protection against crop disease. However, most of the K within the soil exists as a mineral and is not readily available for plant uptake. The K-containing minerals, such as feldspars and micas, are resistant to weathering; this leads to minimal K inputs from weathering during a growing season. Only <" 2 % of soil K is readily available for plant uptake and is either in solution form or exchangeable form (Brady and Weil 2007). Potassium in the soil solution is immediately available to plants. Exchangeable K is a part of the cation exchange capacity of soils and is adsorbed on the soil colloid. The dissolved and exchangeable K concentrations are in equilibrium. For instance, when plants remove K from the soil solution, K is released from the cation exchange complex into the soil solution until K equilibrium is reached. Potassium also occurs in nonexchangeable or fixed form. Fixed K is trapped between clay particles and remains unavailable until it converts to exchangeable K. This conversion usually does not occur within the span of one growing season; therefore, fixed K is considered a slowly available form of K.

Plants uptake large amounts of K in their aboveground biomass, often more than needed for growth. Because of this excessive uptake, harvesting crops or forests can remove large amounts of K from the system each year. Potassium deficiency is normally seen first on the lower plant leaves because

K is translocated from older to younger tissue. Deficiency symptoms include yellow scorching along the leaf edge, slow growth, and weak stalks. Like nitrate, K is extremely mobile and readily lost via leaching. Applying substantial amounts of K fertilizer can lead to excessive plant uptake and loss to leaching. In addition to the soil itself as a natural K source, leaving crop residue on the soil can return a significant portion of K to the soil.

ROLE OF SOIL BIOLOGY

One critical function of soil is to provide a home for organisms. Soil biota plays an integral role in soil ecosystems by decomposing leaves, downed logs, and animals, and also providing the primary nutrient source for vegetation. Soil biota includes both flora (plants) and fauna (animals). Soil fauna subsists on a wide variety of energy sources, including: living plant material (herbivores), animals (carnivores), dead material (detritivores), fungi (fungivores), and bacteria (bacterivores). The size of soil fauna also varies. Macrofauna (> 2 mm) includes animals such as groundhogs, moles, earthworms, centipedes, ants, and termites; mesofauna (0.1–2.0 mm) includes springtails and mites; microfauna (< 0.1 mm) includes species such as rotifers, nematodes, and other single-celled organisms. Soil flora includes organisms as small as diatoms and algae up to the size of tree roots. An important member of soil flora is mycorrhizae, fungi that form a symbiotic (mutually beneficial) relationship with plant roots. Most plants contain roots infected with mycorrhizal fungi. Mycorrhizae enhance water and nutrient absorption by increasing root surface area and accelerate mineral weathering which releases nutrients to the soil (Fisher and Binkley 2000). Collectively, soil biota carries out enzymatic and physical processes that decompose organic matter, build soil humus, and make nutrients available for plants. Table 4 provides a description of the microfauna numbers found in a teaspoon of soil.

Table 4. Brief description of the organisms in a teaspoon of soil. (Source: Melendrez 1975).

Organism	Number of Individuals per Teaspoon of Soil	Primary Role
Bacteria	5 to 500 million in agricultural soils; 20 million to 2 billion in forest soils	Consume readily decomposable materials
Fungi	Up to 40 miles of fungal hyphae	Consume hard to decompose organic matter
Protozoa	100 to 100,000	Feed on bacteria and each other
Nematodes	5 to 500 with high variability among soils	Feed on bacteria, fungi, and plant roots
Microarthropods	Several species	Feed on fungi, plants, and organic

matter

Decomposition is one of the most critical roles that soil biota play in an ecosystem. Without efficient decomposition, organic material would accumulate on the soil surface and nutrients would be bound within the material. Decomposition is initiated immediately when a leaf, twig, or fruit hits the ground. Once on the ground surface, biota begins to physically break down the material, creating more surface area to which flora can adhere.

SOILS AND ROOTS

Roots are a critical component in the soil environment. Plants rely on roots for structure, support, water, and nutrient uptake. The relationship between roots and mycorrhizal fungi increases nutrient availability and absorption. Roots also act as a reservoir for food storage (starches) and sometimes synthesize growth hormones for the plant. Root growth is controlled by soil moisture, compaction, structure, texture, temperature, and chemistry. Once roots decay, the channels left behind improve air and water movement within the soil.

Nutrients are taken up by roots by the processes of root interception, mass flow, and diffusion. Root interception occurs when roots grow towards nutrients in nutrient-rich soil so that they can be utilized by the plant. Since the roots must continually grow in undepleted soil for root interception to occur, this process is limited. Mass flow occurs when nutrients are transported with soil water to a root that is actively extracting water from the soil. This process is most effective in periods of rapid transpiration with high concentrations of nutrients in the soil water solution. Diffusion occurs when nutrients move from high concentration areas (nutrient saturated) to low concentration areas (nutrient depleted) near the root surface. Rates of uptake via diffusion depend on concentration gradients, soil water contents, ion size and charge, soil temperature, and root adsorption rates. Water enters the root through root hairs or the cortex. Osmotic pressure (movement of water from areas of high to low concentration) causes the water to enter into the plant cell. The expansion and contraction of the roots cause water to move up through the cortex, through the xylem vessels, the stem, and into the rest of the plant. A manometer, an instrument used for measuring the pressure of a fluid, is used to measure the pressure of a root system. Root pressure is the pressure exerted on the liquid contents of cortical cells in roots. Cortical cells, which aid in the transport and storage of water and nutrients, are either turgid (expanded) or flaccid (contracted). Turgor pressure is the actual hydrostatic pressure developed inside a cell. This pressure is due to endosmosis, or the inward flow of water from outside of the cell. Flaccidity is when the cell undergoes exosmosis, the opposite of endosmosis, where water is lost and the cell becomes limp. To understand how root pressure works, cut a well-watered plant close to the ground, quickly attach a manometer to the stem and observe how the pressure changes on the gauge. The cut stem will exude water, suggesting pressure from the roots to the stem is being released. Although most water is absorbed through roots, some plants have developed the ability to absorb water through their leaves. Along the California coast, <" 80 % of the redwood species, Sequoia sempervirens, absorb water deposited by fog through their leaves (Limm et al. 2009). This adaptation is especially useful in areas prone to drought, like California.

Plants have both primary roots and secondary roots, which come in many different shapes and sizes. Plant root networks are heterogeneous and no two root systems will be identical. Roots adapt and grow in response to environmental conditions. There are three common root systems of North American tree species. Tap root systems have a prominent tap root and smaller (secondary) lateral roots that grow off the side of the dominant primary root. Tap root systems are effective in areas where access to water is located deep within the soil profile so water can be accessed in periods of drought. Thus, many tree species with tap roots are well-adapted for upland and dry site conditions. Heart root systems occur in both upland and bottomland tree species and are adapted for mesic (moderately moist) site conditions. Flat roots are shallow root systems that often occur in bottomland species. Tree species with flat root systems are commonly uprooted in high windstorms when the ground is near saturation. Contrary to popular belief, <" 90 to 95 % of tree roots are found within 1 m of the soil surface and most tree root systems extend out <" 2 times the width of the crown. Absorption of nutrients occurs mainly in fine roots (< 2 mm in diameter) which are concentrated in the surface horizons (Kimmins 1997). Trees rely on these roots for access to water and nutrients. Root systems of various tree species are shown in Table 5.

Common root system shapes for North American tree species. Drawing by Robin L Quinlivan.

Major features of a root. Drawing by Robin L Quinlivan.

Table 5. Common root systems for various tree species in the United States.

Tap root	Heart Root	Flat root
Hickory	Red Oak	Fir
White Oak	Honey Locust	Spruce
Butternut	Basswood	Sugar Maple
Walnut	Sycamore	Cottonwood
Hornbeam	Pine	Silver Maple

Soil Erosion and Conservation

Soil erosion is a major concern around the globe. In order to properly prevent and manage erosion, it is important to understand erosion concepts. Soil erosion is both naturally-occurring (geologic erosion) and influenced by humans (accelerated erosion). Accelerated erosion can be 10 to 1000 times more damaging than geologic erosion (Brady and Weil 2007). Examples include disturbance of soil or natural vegetation by grazing livestock, deforestation, tillage, and construction, each of which are exacerbated by high rainfall and/or steep slopes.

Globally, <" 15 % of total land area is eroded, of which over half is severely eroded (Blanco-Canqui and Lal 2008). Soil erosion rates are greater than soil formation rates, posing a threat to sustainable agriculture (Pimentel 2006). Productive land is < 11 % of earth's total land area (Eswaran et al. 2001), yet it must supply food to the world's population (> 7 billion people). Pimentel (2006) suggested that soil is being lost at a rate that is 10 to 40 times faster than soil formation and <" 10 million ha of cropland is lost yearly to erosion. Agriculture accounts for <" 75 % of soil erosion worldwide (Pimentel 2006). Soil erosion is lowest in the U.S. and Europe (<" 10 metric tons ha yr soil lost) and much more severe in the rest of the world (<" 30 – 40 metric tons ha yr soil lost) where resources and incentives for implementing conservation measures are inadequate (Blanco-Canqui and Lal 2008).

Erosion is defined by the Soil Science Society of America (SSSA 2008, 19) as: "(i) The wearing away of the land surface by rain or irrigation water, wind, ice, or other natural or anthropogenic agents that abrade, detach, and remove geologic parent material or soil from one point on the earth's surface and deposit it elsewhere, including such processes as gravitational creep and so-called tillage erosion. (ii) The detachment and movement of soil or rock by water, wind, ice, or gravity."

Essentially, erosion occurs in a three-step process of detachment, transport, and deposition. The first step, detachment, is the removal of soil material from the soil mass by raindrops, running water, wind, or human/ animal activities. The detached soil materials are then transported downhill by method of splashing, floating, dragging, or rolling. Sand particles are heavier than silt or clay, and therefore cannot be transported as far or at the same velocity. Silts and clays often become suspended in water and may be transported long distances. The third step, deposition, usually occurs in floodplains, fields, wetlands, lakes/reservoirs, rivers, and streams. In agricultural fields, sediment may also be deposited in front of control structures such as dry dams, terraces, or vegetative filters (e.g., riparian areas and grass waterways).

Soil erosion is divided up into different categories: rill erosion, interrill or sheet erosion, and gully erosion. Rill erosion is the formation of tiny channels across a soil surface that often occurs in sloping agricultural fields. Interrill erosion is the removal of a somewhat uniform layer of soil which is attributed to the splash effect of raindrop impact and also sheet flow. If rill erosion persists, it is likely that gully erosion will occur. Gully erosion is the erosion process that removes large amounts of soil and results in a narrow, deep channel.

Eventually, high amounts of erosion will destroy soil quality, which in turn, compromises long-term food and timber production. The Dust Bowl of the 1930s raised awareness to the need for soil conservation and prompted implementation of conservation practices. Several government programs geared towards soil conservation are in place to minimize and prevent unnaturally high occurrences of soil erosion by wind and water. The Food and Security Act of 1985 included conservation measures that helped reduce erosion from U.S. cropland by 1.3 billion metric tons between 1982 and 1997 (Montgomery 2007). For example, the Conservation Reserve Program (CRP) (included in the Food and Security Act of 1985) provides economic incentives to landowners who implement soil conservation measures on highly erodible cropland. In 2010, 10.7 metric tons ha of cropland was lost in the U.S., whereas CRP land reported a total of 1.7 metric tons ha lost to wind and water erosion (USDA 2013). Since the CRP was implemented in 1986, soil erosion has been reduced by > 7 billion metric tons (USDA 2010). Additionally, recent adoption of notill farming has contributed to a reduction in soil erosion. In a review of studies comparing conventional and notill methods, Montgomery (2007) found that notill practices reduced soil erosion by 2.5 to > 1,000 times compared to conventional methods.

Although adoption of conservation practices is increasing, there are several obstacles that land managers interested in soil conservation must overcome. At times, the initial implementation of conservation practices can be more costly than the economic return. The fear of the unknown will sometimes steer people away from adopting new conservation technologies. It is also difficult to convey the importance of soil conservation to every individual, whether it is due to lack of interest or unavailable information.

The Universal Soil Loss Equation (USLE), a model based on extensive erosion data from small plot studies across the U.S., was created by countless federal and university scientists in the mid-1900s to predict soil loss. The USLE has been used worldwide to estimate soil erosion and guide soil conservation efforts. The USLE equation accounts for slope, rainfall, soil series, topography, crop system, and management practices on a landscape.

The equation has been revised over time and is now referred to as Revised Universal Soil Loss Equation, Version 2 (RUSLE2), a computer-based equation. However, to understand the principles of estimating soil erosion rates, it is important to analyze the original formula:

where A is annual soil loss (tons ac yr), R is erosivity factor, K is erodability factor, LS is slope length and slope steepness factor, C is cover and management factors, and P is conservation practices factor.

The erosivity factor (R) assesses the total rainfall, intensity, and seasonal distribution of rainfall in a geographic location. The R factor accounts for the driving force of rill and interrill erosion. In other words, the R factor recognizes that high intensity rainfall will cause more catastrophic damage than a lower intensity rainfall event. The erodability factor (K) assesses how susceptible the soil is to erosion, which includes a soil's infiltration capacity, texture, and its structural stability/integrity.

Only three of the factors in the equation can be manipulated. The slope length factor (LS) can be changed by reshaping the land into terraces, dry dams, or other erosion control structures. The cover factor (C) is probably the easiest factor to change. In forests, limiting harvesting and road structures can greatly improve this value. In agriculture fields, perennial crops (e.g., alfalfa, orchard grass) that retain groundcover in the winter versus annually harvested crops (e.g., soybeans, corn) will improve the C factor, as will planting cover crops during times the field would normally be fallow. The conservation practices factor (P) can be changed by installing various conservation structures and implementing contour cropping wherever necessary. An example of a conservation structure would be installing a water bar after a timber harvest. Water bars are used to divert water away from logging roads and into vegetated areas, decreasing erosion of these roads.

SOIL QUALITY

The basic definition of soil quality is the capacity of a soil to function (Karlen et al. 1997). This functional definition balances the physical, biological, and chemical components of soil. The expanded version of this functional definition is "the capacity of a specific kind of soil to function, within natural or managed ecosystem boundaries, to sustain plant and animal productivity, maintain or enhance water and air quality, and support human health and habitation" (Karlen et al. 1997). What constitutes good soil quality may be different according to land use and/or geographic region. For this reason, Karlen et al. (1997) suggests a relational approach to evaluating soil quality, as opposed to an absolute approach. Similarly, Doran et al. (1996) suggest that soil quality should be evaluated based on how well a soil functions within its specific ecosystem (agriculture, urban, etc.). For example, in an agricultural field, the capacity of a soil to function at sustaining crop growth would depend on several soil characteristics including bulk density, soil moisture, infiltration, and biological activity, to name a few. Many of these properties can be changed by management (e.g., infiltration and organic matter content) and soil quality can be improved according to its function. As seen throughout this article, soil also affects water quality, air quality, and biotic quality. Protecting and/ or improving soil quality can provide a stepping stone to improving environmental quality as a whole. For example, planting cover crops when a field would otherwise be bare, helps reduce soil erosion and aids in soil and nutrient retention on site, limiting its transportation to waterways where water quality would be affected.

LAND USE AND SOILS

Cultivated and Grazed Soils

In cultivated areas, soils are mechanically worked to break up compaction and amended with fertilizers, herbicides, and pesticides to enhance crop production and provide weed and pest control. These processes can expose bare mineral soil and increase the risk of water and wind erosion. Under improper management, agriculture can devastate the landscape. For example, intense cotton farming in the southeastern United States was blamed for the loss of 10 to 30 cm of topsoil (with an average of <" 18 cm), which led to the accumulation of large volumes of legacy sediment in the floodplains (Trimble 1974). Similarly, poor farming practices coupled with a drought, caused millions of acres of farmland to become lost during the time known as the "dust bowl" or "dirty thirties" era. Fortunately , recent farming practices have adopted environmentally conscientious methods to reduce soil erosion and other impacts on the environment. Farmers now understand that crop selection and farm management decisions should be based on soil characteristics and land conditions. Farmers seek to maximize yields while simultaneously creating a sustainable farming system.

Land use change can have significant impacts on soil quality. If cultivated areas are abandoned, shrubs and trees quickly reoccupy the site; however, carbon pools and nutrient dynamics within the soil may be slow to recover. For example, following a 30-yr succession in North Carolina, most available carbon was allocated to aboveground standing biomass and not towards soil organic matter accumulation (Richter et al. 1995).

Impacts of grazing on soils fall into two categories: 1) physical impacts from animals and 2) chemical and biological impacts from animal waste.

Livestock, such as cattle, compact the soil structure and typically consume vegetation in heavily used areas, such as feeding or watering sites. As a result, the compacted bare areas are more susceptible to erosion and surface runoff. Waste associated with livestock is also a problem. Urine and feces from livestock can create hotspots for nutrients and bacteria. In these areas the vegetative uptake often is much lower than the supply of nutrients. Nitrogen, for instance, either leaves the site through groundwater or surface runoff or is denitrified under optimal conditions. Oftentimes in heavily grazed land, the carbon supply for completed denitrification is lacking, thus incomplete conversion of NOto N gas occurs, resulting in nitrous oxide pollution into the atmosphere.

Forest Soils

Forest soils tend to be on land unsuitable for row-crop agriculture or grazing. Thus, forests are typically in areas of steep topography, poor drainage, or where soils are too rocky for farming. Contrary to agricultural soils, forest soils typically have an intact litter layer that can protect soils from temperature and moisture fluctuations and from wind and water erosion.

Forest Harvesting

Forest harvesting is a common occurrence in the United States and around the world. Harvesting provides the nation with wood products needed in our daily lives. Along with these benefits come potential damages to our environment. The majority of compaction in a forest is caused by forest harvesting procedures. Surprisingly, the primary source of erosion during a harvest results from the construction of roads into the forest and not from the actual logging practices, as one might expect. It is the job of foresters and ecologists to conduct sustainable forest harvesting using best management practices in an effort to reduce site impacts. Harvesting timber has a variety of different effects on the soil, depending on the kind of harvest and caution used while harvesting. The harvesting process opens up the canopy, which in turn allows for more light to hit the forest floor. This increase in light exposure changes the chemical processes occurring in the soil. These changes can then shift the plant and animal species composition found on the site. The exposure of mineral soil from harvesting can increase the potential for erosion and leaching of nutrients.

Loggers use a variety of methods for harvesting timber. Common equipment includes log skidders, feller-bunchers, and cable logging, each with their own pros and cons. Heavy machinery equipped with wide tracks instead of pressured wheels reduces the degree to which the machine compacts the soil. Types of machinery used often depend upon the region. In places like the Midwest where there are large tracts of agricultural land and few tracts of forestland, access to state-of-the-art logging equipment is limited. Detailed guidelines on Best Management Practices (BMPs) associated with forest harvesting are described in "Guiding Principles for Management of Forested, Agricultural, and Urban Watersheds" (Edwards et al. 2015, this issue).

Long-term site productivity continues to be a concern to natural resource managers. An increase in the industries of biofuel production, whole-tree harvesting, woodchip utilization, and wood pellets have researchers and managers studying how long forests can sustain the continuous removal of biomass from forests. A potential short-term solution to this problem could be the introduction of fertilization into forest ecosystems to replenish nutrients exported from the site via whole-tree removal. Large-scale fertilization of forests has increased in the last decade. However, it is unclear how large-scale fertilization affects long-term site productivity because a longer monitoring period is needed to assess the changes in site quality.

Fire

Fire plays a key role in shaping the structure of a forest community; however, fire can also be both beneficial and detrimental to forest soils. Fire can alter the chemical and physical status of soils, impact hydrology, increase soil erosion, and influence the biotic structure of soils. The severity of fire impacts generally depends on burn intensity, climatic conditions following the burn, landscape, and soil characteristics. Following intense burns, nutrient cycling can be disrupted and lead to leaching, volatilization, and/ or transformation. When vegetation is consumed by fire, some of the litter-incorporated N, P, K, Ca, Mg, Cu, Fe, Mn, and Zn are volatilized and lost from the system, while metallic nutrients such as K, Ca, and Mg are converted into oxides and accrued in ash (DeBano et al. 1998). The long-term effects of fire on the soil nutrient status can be significant (Williard et al. 2005). Repeated low intensity fires, such as prescribed fires or single large fires, can reduce soil nitrogen pools significantly (Gagnon 1965; Carreira et al 1994). In fact, Gagnon (1965) documented that a historical fire (20 years earlier) exhibited long-term effects on soil nitrogen and tree nutrition. Vegetation following a 500 km burn showed deficiencies of both nitrogen and phosphorus in foliar analyses.

Using properly executed prescribed fire may actually increase nutrient uptake and provide greater surface roughness through the establishment of dense vegetation post-burn. The increased surface roughness from the newly established and vigorous vegetation growth can reduce surface runoff velocities, in turn reducing the risk of soil erosion. Additionally, the flush of vegetation can increase nutrient uptake, thereby retaining the nutrients on site. Conversely, uncontrolled fire can be detrimental to a forest ecosystem. Under conditions where all vegetation is consumed, the risk of soil erosion is increased, nutrient uptake by plants may be reduced, and high stream temperatures may result. As vegetation is removed, evapotranspiration in the watershed is reduced, thus providing a greater volume of water that can produce surface runoff. Increased overland flow combined with a consumed litter layer can lead to serious erosion problems and result in higher stream discharges. To compound the problem, intense burns can produce hydrophobic conditions in the soil that ultimately reduce infiltration and increase surface runoff. Hydrophobic conditions are generally produced during moderate to high severity burns (175°–200° C), in coarse textured soils, and at lower soil water contents (DeBano et al. 1998). Both nutrient runoff and erosion can be exacerbated in areas of steep slopes and high rainfall.

Soil heating is the primary mechanism by which fires impact a soil's physical and biotic structure (Neary et al. 1999). The degree of soil heating during a fire is highly variable across a site, but depends on fuel characteristics (size, arrangement, and moisture status), soil properties (texture, structure, and both the moisture and organic content), and fire behavior (intensity and duration). Generally 8–10 % of the aboveground heat can be transferred to the soil, resulting in higher soil temperatures (DeBano et al. 1977). Fuels differ in terms of maximum temperatures produced and the duration that they can burn. For example, downed logs and stumps can burn for days following the passage of the fire front, which will induce a more sustained soil heating. Further, finer textured soils (i.e., increased clay content) tend to transfer heat better than quartz-based soils.

Organic content of soils can also be impacted by fire. The consumption of soil organic matter, which is a cementing agent in soils, can result in a loss of soil structure. Reduced structure can reduce porosity, increase bulk density, and decrease infiltration and percolation. Structure takes time to rebuild, and under severe conditions where the soil biota was impacted, the recovery can be even greater since soil invertebrates play a key role in maintaining soil porosity. The degree to which soil invertebrates are impacted relates to soil temperature and moisture. Biota can tolerate higher temperatures in dry soils compared to moist soils.

Urban Soils

Soils are modified by anthropogenic activities in urban environments. The impacts of urban infrastructure on the environment create changes in the physical, chemical, and biological properties of urban soils. Notably, pore space is reduced due to soil compaction from building and road engineering, thereby increasing soil bulk densities. Additionally, the loss of pore space and soil structure combined with increased impervious surfaces leads to reduced infiltration and increased surface runoff. Compaction may also lead to poor establishment of plant communities in urban areas. Pouyat et al. (2007) found that K, P, bulk density, and pH were higher in urban residential soils compared to forest soils, likely a result of lawn fertilizers and intensity of use. However, other soil variables measured were related to parent material rather than land use, including soil texture, Al, Fe, and other micronutrients (Pouyat et al. 2007). While urbanization can impact some soil characteristics, especially bulk density and infiltration, other surface soil characteristics are still predetermined by parent material. Urban soils can recover over time. Scharenbroch et al. (2005) found that as time since the initial disturbance of urban soils increased and allowed for soil development, bulk densities decreased and organic matter and biological activity increased. Time is a soil-forming factor that plays an important role in reducing the impacts of urbanization on soil characteristics.

Knowledge of urban soil properties can lead to better management of water resources in terms of both quantity and quality. Duration of irrigation in urban areas should be based on the water holding capacity of the soil. Organic matter can be incorporated into the soil to increase pore space and soil aggregate stability, thereby increasing infiltration and nutrient retention and reducing runoff. Soil texture should be considered when determining fertilizer management. For instance, sandy soils exhibit poor nutrient retention and nutrients that leach below the rooting zone may eventually travel to ground and surface waterbodies, impacting water quality. Nitrate is especially mobile and nitrate that is not used by plants or microbes is easily transported from the soil water to groundwater.

Index